THE LIBRARY
ST. MARY'S COLLEGE OF MARYLAND
ST. MARY'S CITY, MARYLAND 20686

Undergraduate Texts in Mathematics

Editors
S. Axler
F.W. Gehring
P.R. Halmos

Springer
New York
Berlin
Heidelberg
Barcelona
Budapest
Hong Kong
London
Milan
Paris
Santa Clara
Singapore
Tokyo

Undergraduate Texts in Mathematics

Anglin: Mathematics: A Concise History and Philosophy.
Readings in Mathematics.
Anglin/Lambek: The Heritage of Thales.
Readings in Mathematics.
Apostol: Introduction to Analytic Number Theory. Second edition.
Armstrong: Basic Topology.
Armstrong: Groups and Symmetry.
Axler: Linear Algebra Done Right.
Bak/Newman: Complex Analysis. Second edition.
Banchoff/Wermer: Linear Algebra Through Geometry. Second edition.
Berberian: A First Course in Real Analysis.
Brémaud: An Introduction to Probabilistic Modeling.
Bressoud: Factorization and Primality Testing.
Bressoud: Second Year Calculus.
Readings in Mathematics.
Brickman: Mathematical Introduction to Linear Programming and Game Theory.
Browder: Mathematical Analysis: An Introduction.
Buskes/van Rooij: Topological Spaces: From Distance to Neighborhood.
Cederberg: A Course in Modern Geometries.
Childs: A Concrete Introduction to Higher Algebra. Second edition.
Chung: Elementary Probability Theory with Stochastic Processes. Third edition.
Cox/Little/O'Shea: Ideals, Varieties, and Algorithms. Second edition.
Croom: Basic Concepts of Algebraic Topology.
Curtis: Linear Algebra: An Introductory Approach. Fourth edition.
Devlin: The Joy of Sets: Fundamentals of Contemporary Set Theory. Second edition.
Dixmier: General Topology.
Driver: Why Math?
Ebbinghaus/Flum/Thomas: Mathematical Logic. Second edition.
Edgar: Measure, Topology, and Fractal Geometry.
Elaydi: Introduction to Difference Equations.
Exner: An Accompaniment to Higher Mathematics.
Fine/Rosenberger: The Fundamental Theorem of Algebra.
Fischer: Intermediate Real Analysis.
Flanigan/Kazdan: Calculus Two: Linear and Nonlinear Functions. Second edition.
Fleming: Functions of Several Variables. Second edition.
Foulds: Combinatorial Optimization for Undergraduates.
Foulds: Optimization Techniques: An Introduction.
Franklin: Methods of Mathematical Economics.
Gordon: Discrete Probability.
Hairer/Wanner: Analysis by Its History.
Readings in Mathematics.
Halmos: Finite-Dimensional Vector Spaces. Second edition.
Halmos: Naive Set Theory.
Hämmerlin/Hoffmann: Numerical Mathematics.
Readings in Mathematics.
Hijab: Introduction to Calculus and Classical Analysis.
Hilton/Holton/Pedersen: Mathematical Reflections: In a Room with Many Mirrors.
Iooss/Joseph: Elementary Stability and Bifurcation Theory. Second edition.
Isaac: The Pleasures of Probability.
Readings in Mathematics.
James: Topological and Uniform Spaces.
Jänich: Linear Algebra.
Jänich: Topology.

(continued following index)

Benjamin Fine   Gerhard Rosenberger

The Fundamental Theorem of Algebra

 Springer

Benjamin Fine
Department of Mathematics
Fairfield University
Fairfield, CT 06430
USA

Gerhard Rosenberger
Department of Mathematics
University of Dortmund
Dortmund
Germany

Editorial Board

S. Axler
Department of
  Mathematics
Michigan State University
East Lansing, MI 48824
USA

F.W. Gehring
Department of
  Mathematics
University of Michigan
Ann Arbor, MI 48109
USA

P.R. Halmos
Department of
  Mathematics
Santa Clara University
Santa Clara, CA 95053
USA

Mathematics Subject Classification (1991): 11-01, 30-01, 54-01

Library of Congress Cataloging-in-Publication Data
Fine, Benjamin, 1948–
  The fundamental theorem of algebra / Benjamin Fine,
  Gerhard Rosenberger.
    p. cm. — (Undergraduate texts in mathematics)
  Includes bibliographical references and index.
  ISBN 0-387-94657-8 (hardcover : alk. paper)
  1. Fundamental theorem of algebra. I. Rosenberger, Gerhard.
II. Title. III. Series.
QA212.F55 1997
512.9'42—dc21   96-53013

Printed on acid-free paper.

© 1997 Springer-Verlag New York, Inc.
All rights reserved. This work may not be translated or copied in whole or in part without the written permission of the publisher (Springer-Verlag New York, Inc., 175 Fifth Avenue, New York, NY 10010, USA), except for brief excerpts in connection with reviews or scholarly analysis. Use in connection with any form of information storage and retrieval, electronic adaptation, computer software, or by similar or dissimilar methodology now known or hereafter developed is forbidden. The use of general descriptive names, trade names, trademarks, etc., in this publication, even if the former are not especially identified, is not to be taken as a sign that such names, as understood by the Trade Marks and Merchandise Marks Act, may accordingly be used freely by anyone.

Production managed by Timothy Taylor; manufacturing supervised by Jeffrey Taub.
Camera-ready copy prepared from the author's files by the Bartlett Press, Inc.
Printed and bound by Maple-Vail Book Manufacturing Group, York, PA.
Printed in the United States of America.

9 8 7 6 5 4 3 2 1

ISBN 0-387-94657-8 Springer-Verlag New York Berlin Heidelburg SPIN 10524234

To our families:

>Linda, Carolyn, and David
>Katariina, Anja, and Aila

Preface

These notes grew out of two courses, one given in the United States and one given in Germany on the Fundamental Theorem of Algebra. The purpose of these courses was to present a great deal of nonelementary mathematics, all centered on a single topic. The Fundamental Theorem of Algebra was ideal for this purpose. Analysis, algebra and topology each have developed different techniques which surround this result. These techniques lead to different proofs and different views of this important result. It is startling how much mathematics can be introduced and learned in this manner.

In the United States it was presented as a "capstone" course for upper level undergraduates. Many of the topics were familiar to the students but many were new. The goal of continually returning to a proof of the Fundamental Theorem of Algebra gave a focus to a large body of (what is at first glance) seemingly unrelated material. In addition, many nice applications, such as the insolvability of the quintic and the transcendence of e and π could be introduced. We feel that undergraduates in such a capstone course are an ideal audience for the book. Many departments in the U.S. are adopting the idea of a summary course. In addition, the book could serve as a foundation reference for beginning graduate students. We also feel that the algebra sections, Chapters 2, 3, 6, 7, could be used, with some additions from outside sources, as an alternative version of an undergraduate algebra course or as a supplement for such a course. The United States version of the course covered in one semester, with some omissions, most of the material in Chapters 1 through 7. The whole book could be covered at a relatively moderate pace in two semesters.

In Germany the material was presented to a class of potential teachers. A high school (or in Germany, *gymnasium*) teacher should be exposed to a wide range of mathematical topics. This material fulfilled this objective for this audience. It is our hope that similar teacher training courses in the U.S. might also adopt these notes. In the course in Germany, essentially the whole book was presented in two semesters.

We wish to thank Nicole Isermann for her extremely careful proofreading of the manuscript. We also wish to thank Kati Bencsath and Bruce Chandler for reading preliminary versions and making suggestions, and finally, we would like to thank Paul Halmos for his helpful suggestions.

Benjamin Fine, *Fairfield University, United States*
Gerhard Rosenberger, *Universität Dortmund, Germany*

Contents

	Preface	vii
1	**Introduction and Historical Remarks**	1
2	**Complex Numbers**	5
	2.1 Fields and the Real Field	5
	2.2 The Complex Number Field	10
	2.3 Geometrical Representation of Complex Numbers	12
	2.4 Polar Form and Euler's Identity	14
	2.5 DeMoivre's Theorem for Powers and Roots	17
	Exercises	19
3	**Polynomials and Complex Polynomials**	21
	3.1 The Ring of Polynomials over a Field	21
	3.2 Divisibility and Unique Factorization of Polynomials	24
	3.3 Roots of Polynomials and Factorization	27
	3.4 Real and Complex Polynomials	29
	3.5 The Fundamental Theorem of Algebra: Proof One	31
	3.6 Some Consequences of the Fundamental Theorem	33
	Exercises	34
4	**Complex Analysis and Analytic Functions**	36
	4.1 Complex Functions and Analyticity	36

	4.2 The Cauchy–Riemann Equations	41
	4.3 Conformal Mappings and Analyticity	46
	Exercises	49

5 Complex Integration and Cauchy's Theorem — 52

    5.1 Line Integrals and Green's Theorem — 52
    5.2 Complex Integration and Cauchy's Theorem — 61
    5.3 The Cauchy Integral Formula and Cauchy's Estimate — 66
    5.4 Liouville's Theorem and the Fundamental Theorem of Algebra: Proof Two — 70
    5.5 Some Additional Results — 71
    5.6 Concluding Remarks on Complex Analysis — 72
    Exercises — 72

6 Fields and Field Extensions — 74

    6.1 Algebraic Field Extensions — 74
    6.2 Adjoining Roots to Fields — 81
    6.3 Splitting Fields — 84
    6.4 Permutations and Symmetric Polynomials — 86
    6.5 The Fundamental Theorem of Algebra: Proof Three — 91
    6.6 An Application—The Transcendence of e and π — 94
    6.7 The Fundamental Theorem of Symmetric Polynomials — 99
    Exercises — 102

7 Galois Theory — 104

    7.1 Galois Theory Overview — 104
    7.2 Some Results From Finite Group Theory — 105
    7.3 Galois Extensions — 112
    7.4 Automorphisms and the Galois Group — 115
    7.5 The Fundamental Theorem of Galois Theory — 119
    7.6 The Fundamental Theorem of Algebra: Proof Four — 123
    7.7 Some Additional Applications of Galois Theory — 124
    7.8 Algebraic Extensions of $\mathbb{R}$ and Concluding Remarks — 130
    Exercises — 132

8 Topology and Topological Spaces — 134

    8.1 Winding Number and Proof Five — 134
    8.2 Topology—An Overview — 136
    8.3 Continuity and Metric Spaces — 138
    8.4 Topological Spaces and Homeomorphisms — 144
    8.5 Some Further Properties of Topological Spaces — 146
    Exercises — 149

9	**Algebraic Topology and the Final Proof**	**152**
	9.1 Algebraic Topology	152
	9.2 Some Further Group Theory—Abelian Groups	154
	9.3 Homotopy and the Fundamental Group	159
	9.4 Homology Theory and Triangulations	166
	9.5 Some Homology Computations	173
	9.6 Homology of Spheres and Brouwer Degree	176
	9.7 The Fundamental Theorem of Algebra: Proof Six	178
	9.8 Concluding Remarks	180
	Exercises	180
	Appendix A: A Version of Gauss's Original Proof	**182**
	Appendix B: Cauchy's Theorem Revisited	**187**
	Appendix C: Three Additional Complex Analytic Proofs of the Fundamental Theorem of Algebra	**195**
	Appendix D: Two More Topological Proofs of the Fundamental Theorem of Algebra	**199**
	Bibliography and References	**202**
	Index	**205**

CHAPTER 1

Introduction and Historical Remarks

The **Fundamental Theorem of Algebra** states that any complex polynomial must have a complex root. This basic result, whose first accepted proof was given by Gauss, lies really at the intersection of the theory of numbers and the theory of equations, and arises also in many other areas of mathematics. The purpose of these notes is to examine three pairs of proofs of the theorem. The first proof in each pair is fairly straightforward and depends only on what could be considered elementary mathematics. However, each of these first proofs lends itself to generalizations that in turn lead to more general results from which the Fundamental Theorem can be deduced as a direct consequence. These general results constitute the second proof in each pair.

Recall that a **complex polynomial** is a complex function of the form

$$P(z) = a_n z^n + a_{n-1} z^{n-1} + \cdots + a_0,$$

where $a_0, a_1, \ldots, a_n$ are complex numbers and n is a natural number. A **root**, or **zero**, of this polynomial is a complex number z_0 such that $P(z_0) = 0$.

The reasons for the different proofs of this result are due to the distinct characteristics of complex polynomials. First of all, complex polynomials are complex functions, that is, functions from $\mathcal{C}$ to $\mathcal{C}$. As with real polynomials, complex polynomials are everywhere differentiable and so in the language of complex analysis are part of the class of **entire functions**. In this context the Fundamental Theorem of Algebra is a direct consequence of a general result called **Liouville's theorem**. This result states that an entire function that is bounded in the complex plane must be a constant. In Chapters 2 and 3 we introduce the basic results on complex

1

numbers and complex polynomials. We then use these to present a proof of the Fundamental Theorem that utilizes only advanced calculus. This proof suggests Liouville's theorem. In Chapters 4 and 5 we then present the results from complex function theory – specifically complex differentiation, analytic functions, complex integration and Cauchy's theorem – needed to derive Liouville's theorem. From this we give our second proof of the Fundamental Theorem of algebra.

In a different direction, a complex polynomial is an algebraic object. In this context the Fundamental Theorem of Algebra can be phrased as, "the complex numbers are algebraically closed." In Chapter 6 we present the results concerning construction of field extensions and then present a proof of the Fundamental Theorem that depends only on the facts that odd-degree real polynomials have real roots and that given an irreducible polynomial $f(x)$ over a field F, a field extension $F*$ of F can be constructed such that $f(x)$ has a root in $F*$. This proof suggests the following generalization. If K is a field where odd-degree polynomials have roots and $i = \sqrt{-1}$, then $K(i)$ is algebraically closed. The proof of this generalization involves **Galois theory**. In Chapter 7 we present the basic results on group theory and Galois theory needed to understand this proof.

Finally, a complex polynomial is a topological mapping. If we adjoin the point at infinity to the complex plane we obtain a sphere, the **Riemann sphere** S^2. Since $P(\infty) = \infty$ for any complex polynomial $P(z)$, $P(z)$ can be considered as a continuous mapping $P : S^2 \to S^2$. Such topological mappings have what is termed a **winding number**, indicating how much the image of a curve C^1 on S^2 winds around when mapped to S^2. In a similar manner the function $f(z) = z^n$ winds the complex number z around the origin. In Chapter 8 we first present a proof of the Fundamental Theorem using the winding properties of $f(z) = z^n$. This is then generalized to winding numbers of functions $S^2 \to S^2$, from which the Fundamental Theorem is re-obtained. To handle this last generalization we must introduce some basic ideas and techniques in both point-set topology and algebraic topology. This is done Chapters 8 and 9. This final proof requires the most development and is therefore the least self-contained.

There are many variations of the proofs that we present. In a series of appendices we give six additional proofs, each somewhat different from those given in the main body of the notes. In Appendix A we give a modern version of Gauss's original first proof (see below). In Appendix C we present three additional proofs arising out of complex analysis. These require a more detailed analysis of Cauchy's theorem than the one given in Chapter 5. This analysis is given in Appendix B. Finally, in Appendix D we give two addditional topologically motivated proofs. These also depend on the concept of winding number but differ from the two given in Chapters 8 and 9.

We suppose that the reader has been introduced to advanced calculus as least as far as Green's theorem; has studied some abstract algebra, in

1. Introduction and Historical Remarks

particular the definitions of groups, rings and fields, and has studied some linear algebra, in particular the abstract definition of a vector space over a general field. Beyond this we have tried to make these notes as self-contained as possible. However, our goal is to arrive at an understanding of the proofs of the Fundamental Theorem. Therefore, along the way we have proved only those results that can be obtained directly and have left more difficult results (such as the proof of the fundamental theorem of Galois theory) to the references. A note here about terminology. In standard usage equations have roots, and functions and polynomials have zeros. Historically however, the word *root* has been most often connected with the Fundamental Theorem of Algebra. Therefore we use the term *root* throughout these notes, and do not really differentiate between the zero of the polynomial $P(z)$ and the root of the polynomial equation $P(z) = 0$.

The first mention of the Fundamental Theorem of Algebra, in the form that every polynomial equation of degree n has exactly n roots, was given by Peter Roth of Nurnberg in 1608. However its conjecture is generally credited to Girard who also stated the result in 1629. It was then more clearly stated by Descartes in 1637 who also distinguished between real and imaginary roots. The first published proof of the Fundamental Theorem of Algebra was then given by D'Alembert in 1746. However there were gaps in D'Alembert's proof and the first fully accepted proof was that given by Gauss in 1797 in his Ph.D. thesis. This was published in 1799. Interestingly enough, in reviewing Gauss's original proof, modern scholars tend to agree that there are as many holes in this proof as in D'Alembert's proof. Gauss, however, published three other proofs with no such holes. He published second and third proofs in 1816 while his final proof, which was essentially another version of the first, was presented in 1849.

In the main part of these notes we do not touch on Gauss's original proof, which in outline went as follows. Since $P(z)$ is a complex number for any $z \in \mathcal{C}$ and since $z = x + iy$ with $x, y \in \mathbb{R}$ we have $P(z) = u(x, y) + iv(x, y)$. The equations $u(x, y) = 0$ and $v(x, y) = 0$ then represent curves in the plane $\mathbb{R}^2$. By a careful examination of the possible functions $u(x, y), v(x, y)$ for a complex polynomial $P(z)$, Gauss showed that the curves $u(x, y) = 0, v(x, y) = 0$ must have a common solution (x_0, y_0). The complex number $z_0 = x_0 + iy_0$ is then a root of $P(z)$. In Appendix A we present a version of Gauss's original proof.

The Fundamental Theorem of Algebra is actually part of a general development in the theory of equations. The ability to solve quadratic equations and in essence the quadratic formula was known to the Babylonians some 3600 years ago. With the discovery of imaginary numbers, the quadratic formula then says that any degree two polynomial over $\mathcal{C}$ has a root in $\mathcal{C}$. In the sixteenth century the Italian mathematician Niccolo Tartaglia discovered a similar formula in terms of radicals to solve cubic equations. This **cubic formula** is now known erroneously as **Cardano's**

formula in honor of Cardano, who first published it in 1545. An earlier special version of this formula was discovered by Scipione del Ferro. Cardano's student Ferrari extended the formula to solutions by radicals for fourth degree polynomials. The combination of these formulas says that complex polynomials of degree four or less must have complex roots. In the seventeenth century it became clear from the elementary properties of continuous functions that all odd degree real polynomials must have real roots. All these results lent credence to the Fundamental Theorem which was mentioned by Roth in 1608 and conjectured by Girard in 1629.

From Cardano's work until the very early nineteenth century, attempts were made to find similar formulas for degree five polynomials. In 1805 Ruffini proved that fifth degree polynomial equations are insolvable by radicals in general. Therefore there exists no comparable formula for degree 5. Abel in 1825 - 1826 and Galois in 1831 extended Ruffini's result and proved the insolubility by radicals for all degrees five or greater. In doing this, Galois developed a general theory of field extensions and its relationship to group theory. This has come to be known as **Galois theory** (which we will discuss in Chapter 7). More about the history of the Fundamental Theorem of Algebra can be be found in the article by R. Remmert [R].

2

CHAPTER

Complex Numbers

2.1 Fields and the Real Field

Recall that a **field** F is a set with two binary operations, addition, denoted by +, and multiplication denoted by · or just by juxtaposition, defined on it satisfying the following nine axioms:

(1) Addition is commutative: $a + b = b + a$ for each pair a, b in F.
(2) Addition is associative: $a + (b + c) = (a + b) + c$ for $a, b, c \in F$.
(3) There exists an additive identity, denoted by 0, such that $a + 0 = a$ for each $a \in F$.
(4) For each $a \in F$ there exists an additive inverse denoted $-a$, such that $a + (-a) = 0$.
(5) Multiplication is associative: $a(bc) = (ab)c$ for $a, b, c \in F$.
(6) Multiplication is distributive over addition: $a(b + c) = ab + ac$ for $a, b, c \in F$.
(7) Multiplication is commutative: $ab = ba$ for each pair a, b in F.
(8) There exists an multiplicative identity denoted by 1 (not equal to 0) such that $a1 = a$ for each a in F.
(9) For each $a \in F$, with $a \neq 0$ there exists a multiplicative inverse denoted by a^{-1}, such that $aa^{-1} = 1$.

A set G with one operation, +, on it satisfying axioms (1) through (4) is called an **abelian group**. Axioms (1) through (6) define a **ring**, while (1) through (8) define a **commutative ring with an identity.** Therefore, in a more general algebraic context a field can be defined as a commutative ring with an identity in which every nonzero element has a multiplicative inverse.

A field can be considered as the most basic algebraic structure in which all the arithmetic operations: addition, subtraction (addition of additive inverses), multiplication, and division (multiplication by multiplicative inverses) can be done.

Examples of fields include the **rational numbers** $\mathbb{Q}$, the **integers modulo any prime p**, denoted by $\mathbb{Z}_p$, and the field of **real numbers** $\mathbb{R}$. Another example of a field, which indicates the type of algebraic extensions that we will be looking at later on, is the following:

EXAMPLE 2.1.1
Consider the set

$$\mathbb{Q}(\sqrt{2}) = \{a + b\sqrt{2}; a, b \in \mathbb{Q}\}.$$

On $\mathbb{Q}(\sqrt{2})$ define addition and multiplication via algebraic manipulation; that is, if $x = a + b\sqrt{2}, y = c + d\sqrt{2}$ then

$$x + y = (a + c) + (b + d)\sqrt{2}$$

and

$$xy = (ac + 2bd) + (bc + ad)\sqrt{2}.$$

To verify that $\mathbb{Q}(\sqrt{2})$ is a field we must show that is satisfies properties (1) through (9). We leave most of this to the exercises and only show that nonzero elements have multiplicative inverses.

First of all we can identify $\mathbb{Q}$ with $\{a + 0\sqrt{2}; a \in \mathbb{Q}\}$ so that $\mathbb{Q}$ can be considered as a subset of $\mathbb{Q}(\sqrt{2})$. From the definitions of addition and multiplication it is clear that $0 \in \mathbb{Q}$ and $1 \in \mathbb{Q}$ are respectively the additive and multiplicative identities in $\mathbb{Q}(\sqrt{2})$. Now suppose $x = a + b\sqrt{2}$ with $x \neq 0$. Then a and b are not both zero and since $\sqrt{2}$ is irrational, $a^2 - 2b^2 \neq 0$. Let $\bar{x} = a - b\sqrt{2}$. Then by a straightforward computation we have

$$x\bar{x} = a^2 - 2b^2.$$

Now let

$$y = \frac{\bar{x}}{a^2 - 2b^2} = \frac{a}{a^2 - 2b^2} - \frac{b\sqrt{2}}{a^2 - 2b^2}.$$

The value y is well-defined since $a^2 - 2b^2 \neq 0$. By computation it follows that $xy = 1$ and hence y is the multiplicative inverse of x.

This example can be generalized to $\mathbb{Q}(\sqrt{d})$, where d is any rational number that is not a perfect square. □

If F_1 is a subset of a field F that is also a field under the same operations and identities as F we say that F_1 is a **subfield** of F and F is an **extension**

2.1. Fields and the Real Field

field of F_1. As examples we have that $\mathbb{Q}$ is a subfield of $\mathbb{Q}(\sqrt{2})$ and also a subfield of $\mathbb{R}$.

If F_1 is a nonempty subset of a field F, then it will automatically satisfy the associative, commutative, and distributive properties. Therefore, it will be a subfield if it is closed under addition and multiplication, contains 0 and 1, and contains the additive and multiplicative inverses of any element in it. We can summarize this by saying that it must be closed under addition, subtraction, multiplication and division, in which case it will automatically contain 0 and 1.

Theorem 2.1.1
$F_1 \subseteq F$ is a subfield if and only if $F_1 \neq \emptyset$ and F_1 is closed under addition, subtraction, multiplication, and division.

EXAMPLE 2.1.2
Show that $\mathbb{Q}(\sqrt{2})$ is a subfield of $\mathbb{R}$ and therefore a field.

In example 2.1.1 and in the exercises we showed directly that $\mathbb{Q}(\sqrt{2})$ is a field. Here we use Theorem 2.1.1 to accomplish the same thing.

Now, $\mathbb{Q} \subset \mathbb{Q}(\sqrt{2})$, so $\mathbb{Q}(\sqrt{2}) \neq \emptyset$. Therefore, to show that it is a field we must show that $\mathbb{Q}(\sqrt{2})$ closed under the four arithmetic operations. Suppose then that $x = a + b\sqrt{2}$ and $y = c + d\sqrt{2}$ with $y \neq 0$ and $a, b, c, d \in \mathbb{Q}$. Then:

$$x \pm y = (a \pm c) + (b \pm d)\sqrt{2} \in \mathbb{Q}(\sqrt{2})$$

$$\text{since } a \pm c \in \mathbb{Q}, b \pm d \in \mathbb{Q}$$

$$xy = (ac + 2bd) + (bc + ad)\sqrt{2} \in \mathbb{Q}(\sqrt{2})$$

$$\text{since } ac + 2bd \in \mathbb{Q}, bc + ad \in \mathbb{Q}$$

$$1/y = \frac{c}{c^2 - 2d^2} - \frac{d\sqrt{2}}{c^2 - 2d^2} \in \mathbb{Q}(\sqrt{2})$$

$$\text{since } \frac{c}{c^2 - 2d^2} \in \mathbb{Q}, \frac{d}{c^2 - 2d^2} \in \mathbb{Q}$$

From the last two facts it follows that x/y is in $\mathbb{Q}(\sqrt{2})$. Therefore $\mathbb{Q}(\sqrt{2})$ is nonempty and closed under the arithmetic operations so it is a subfield of $\mathbb{R}$. □

Recall that a **vector space** V over a field F consists of an abelian group V together with scalar multiplication from F satisfying:

(1) $fv \in V$ if $f \in F, v \in V$.
(2) $f(u + v) = fu + fv$ for $f \in F, u, v \in V$.
(3) $(f + g)v = fv + gv$ for $f, g \in F, v \in V$.
(4) $(fg)v = f(gv)$ for $f, g \in F, v \in V$.

(5) $1v = v$ for $v \in V$.

If F_1 is a subfield of F then multiplication of elements of F by elements of F_1 are still in F. Since F is an abelian group under addition, F can be considered as a vector space over F_1. Thus any extension field is a vector space over any of its subfields. If F is an extension field of F_1, the dimension of F, as a vector space over F_1, is called the **degree of the extension**.

EXAMPLE 2.1.3

(1) $\mathbb{Q}(\sqrt{2})$ has degree 2 over $\mathbb{Q}$ since $\{1, \sqrt{2}\}$ form a basis.
(2) $\mathbb{R}$ is infinite dimensional over $\mathbb{Q}$, so $\mathbb{R}$ is an extension of infinite degree over $\mathbb{Q}$. □

This second fact depends on the existence of **transcendental numbers**. An element $r \in \mathbb{R}$ is **algebraic** (over $\mathbb{Q}$) if it satisfies some nonzero polynomial with coefficients from $\mathbb{Q}$. That is, $P(r) = 0$, where

$$0 \neq P(x) = a_0 + a_1 x + \cdots + a_n x^n \text{ with } a_i \in \mathbb{Q}.$$

Any $q \in \mathbb{Q}$ is algebraic since if $P(x) = x - q$ then $P(q) = 0$. However, many irrationals are also algebraic. For example $\sqrt{2}$ is algebraic since $x^2 - 2 = 0$ has $\sqrt{2}$ as a root. An element $r \in \mathbb{R}$ is **transcendental** if it is not algebraic.

In general it is very difficult to show that a particular element is transcendental. However there are uncountably many transcendental elements (see exercises). Specific examples are our old friends e and π. We give a proof of their transcendence in Chapter 6.

Since e is transcendental, for any natural number n the set of vectors $\{1, e, e^2, \ldots, e^n\}$ must be independent over $\mathbb{Q}$, for otherwise there would be a polynomial that e would satisfy. Therefore, we have infinitely many independent vectors in $\mathbb{R}$ over $\mathbb{Q}$ which would be impossible if $\mathbb{R}$ had finite degree over $\mathbb{Q}$.

We will return to these ideas later on. Now we must take a closer look at the real field $\mathbb{R}$, which satisfies certain special properties that are essential for the Fundamental Theorem of Algebra.

First of all $\mathbb{R}$, is an **ordered field**. By this we mean that there exists a set of positive reals $\mathbb{R}^+$ closed under addition and multiplication and such that $\mathbb{R}$ satisfies the **trichotomy law**: if $r \in \mathbb{R}$ then either $r \in \mathbb{R}^+$ or $r = 0$ or $-r \in \mathbb{R}^+$ and only one of these holds. This allows us to define an order on $\mathbb{R}$ by $x > y$ if $x - y \in \mathbb{R}^+$.

Being an ordered field forces the square of any nonzero element to be positive: if $r \in \mathbb{R}^+$ then $r^2 \in \mathbb{R}^+$, while if $-r \in \mathbb{R}^+$ then $r^2 = (-r)^2 \in \mathbb{R}^+$. Therefore, only positive elements of $\mathbb{R}$ can have squareroots. In particular, $\sqrt{-1}$ does not exist in $\mathbb{R}$.

2.1. Fields and the Real Field

EXAMPLE 2.1.4
$\mathbb{Q}$, as well as any other subfield of $\mathbb{R}$, is also an ordered field. However, the integers modulo a prime, $\mathbb{Z}_p$, is not. Having a set of positive elements closed under addition and multiplication and satisfying the trichotomy law forces a field to be infinite. Therefore, ordered fields must be infinite, and since $\mathbb{Z}_p$ is finite it cannot be an ordered field. □

Next, $\mathbb{R}$ is **Archimedean**. Given $x, y \in \mathbb{R}$ there exists a natural number n such that $nx > y$. $\mathbb{Q}$ is also Archimedean.

Finally, $\mathbb{R}$ is **complete** in that it satisfies the **least upper bound property, (LUB property)**.

LUB Property
If $S \subset \mathbb{R}$ is bounded above, then S has a **least upper bound** s_o ($s \leq s_o$ for all $s \in S$, and if $s \leq s_1$ for all $s \in S$ then $s_o \leq s_1$).

The LUB property is equivalent to either of two other properties that are sometimes taken as the definition of completeness for $\mathbb{R}$.

Cauchy Sequence Property
If $\{x_n\}$ is a Cauchy sequence in $\mathbb{R}$, then $\{x_n\}$ must converge. Recall that a **Cauchy sequence** is a sequence $\{x_n\}$ such that for all $\epsilon > 0$ there exists N such that $|x_n - x_m| < \epsilon$ for all $m, n > N$.

Nested Intervals Property
If $\{I_n\}$ is a sequence of nested closed intervals ($I_{n+1} \subset I_n$) whose lengths go to zero then there exists a unique point x_o common to all the intervals.

Notice that $\mathbb{Q}$ is *not* complete. Consider the set $S = \{q \in \mathbb{Q}; q^2 < 2\}$. Then S is bounded above by 3 in $\mathbb{Q}$. However, S has no lub in $\mathbb{Q}$ for if it did, it would have to agree with the lub in $\mathbb{R}$ which is the irrational number $\sqrt{2}$.

The completeness property for $\mathbb{R}$ allows us to prove the **intermediate value theorem** for continuous functions $f : \mathbb{R} \to \mathbb{R}$.

Theorem 2.1.2 (Intermediate Value Theorem)
Let $f:[a,b] \to \mathbb{R}$ be a continuous function. If $f(a) < k < f(b)$ or $f(a) > k > f(b)$, then there exists $c \in (a,b)$ with $f(c) = k$. (A continuous function takes any value between any two of its values).

Important for the Fundamental Theorem of Algebra is the following corollary.

Corollary 2.1.1
Suppose $f:[a,b] \to \mathbb{R}$ is continuous and $f(a)f(b) < 0$. Then there exists a value c with $a < c < b$ and $f(c) = 0$.

To see the corollary notice that $f(a)f(b) < 0$ implies that $f(a)$ and $f(b)$ have different signs. Without loss of generality suppose $f(a) < 0, f(b) > 0$. Then 0 is between $f(a)$ and $f(b)$, and the intermediate value theorem applies.

These special properties completely determine $\mathbb{R}$ up to **isomorphism.** If R and S are rings, a function $f : R \to S$ is a **homomorphism** if it satisfies:

(1) $f(r_1 + r_2) = f(r_1) + f(r_2)$ for $r_1, r_2 \in R$.
(2) $f(r_1 r_2) = f(r_1)f(r_2)$ for $r_1, r_2 \in R$.

If f is also a bijection, then f is an **isomorphism**, and R and S are **isomorphic**. Isomorphic algebraic structures are essentially algebraically the same. We have the following theorem.

Theorem 2.1.3
Let F be a complete Archimedean ordered field. Then F is isomorphic to $\mathbb{R}$.

2.2 The Complex Number Field

In the real numbers $\mathbb{R}$, $\sqrt{-1}$ does not exist. Now we formally define $\mathbf{i} = \sqrt{-1}$; that is $\mathbf{i}$ is a new element such that $\mathbf{i}^2 = -1$. Historically, $\mathbf{i}$ was called the **imaginary unit**, but as we will see in the next section, $\mathbf{i}$ has a very real geometric significance.

A **complex number** is then an expression of the form $z = x + \mathbf{i} y$ with $x, y \in \mathbb{R}$. If $x = 0, y \neq 0$, so that z has the form $\mathbf{i} y$, then z is called a (purely) **imaginary number**. The set of **complex numbers**, denoted by $\mathbb{C}$, is then

$$\mathbb{C} = \{x + \mathbf{i} y; x, y \in \mathbb{R}\}.$$

If we identify a real number x with the complex number $x + 0\,\mathbf{i}$, we see that $\mathbb{R} \subset \mathbb{C}$ as sets.

On $\mathbb{C}$ we define arithmetic by algebraic manipulation using the fact that $\mathbf{i}^2 = -1$. That is, if $z = x + \mathbf{i} y, w = a + \mathbf{i} b$ then:

(i) $z \pm w = (x \pm a) + \mathbf{i}(y \pm b)$.
(ii) $zw = (xa - yb) + \mathbf{i}(xb + ya)$.

$$((x + \mathbf{i} y)(a + \mathbf{i} b) = xa + \mathbf{i}(ya) + \mathbf{i}(xb) + \mathbf{i}^2(yb) = (xa - yb) + \mathbf{i}(xb + ya)).$$

2.2. The Complex Number Field

EXAMPLE 2.2.1
Let $z = 3 + 4\mathbf{i}$, $w = 7 - 2\mathbf{i}$ then :

(1) $z + w = 10 + 2\mathbf{i}$.
(2) $z - w = -4 + 6\mathbf{i}$.
(3) $zw = 29 + 22\mathbf{i}$. □

It is easy to verify [see the exercises] that under these definitions $\mathbb{C}$ forms a commutative ring with an identity and that $\mathbb{R}$ is a subring of $\mathbb{C}$. The zero is $0 \in \mathbb{R}$, identified as $0 + 0\mathbf{i}$, and the multiplicative identity is $1 \in \mathbb{R}$ identified with $1 + 0\mathbf{i}$. In order for $\mathbb{C}$ to be a field we must have multiplicative inverses. We now show how to construct these.

Definition 2.2.1
If $z \in \mathbb{C}$ with $z = x + \mathbf{i}y$, then the **complex conjugate** of z, denoted by $\bar{z}$, is $\bar{z} = x - \mathbf{i}y$, and the **absolute value**, or **modulus**, of z, denoted by $|z|$, is $|z| = \sqrt{x^2 + y^2}$.

EXAMPLE 2.2.2
Let $z = 3 + 4\mathbf{i}$. Then $\bar{z} = 3 - 4\mathbf{i}$ and $|z| = \sqrt{9 + 16} = 5$. □

The following lemmas give the properties of the complex conjugate and the absolute value, and from these we will be able to construct inverses.

Lemma 2.2.1
Let $z,w \in \mathbb{C}$. Then

(1) $|z| \geq 0$, and $|z| = 0$ if and only if $z = 0$.
(2) $|zw| = |z||w|$.
(3) $|z + w| \leq |z| + |w|$.

The proofs of these are straightforward computations and are left to the exercises. Notice that these are precisely the properties of the absolute value of real numbers, and so they will allow us to do calculus on $\mathbb{C}$.

Lemma 2.2.2
Let $z,w \in \mathbb{C}$. Then

(1) $\overline{z + w} = \bar{z} + \bar{w}$.
(2) $\overline{zw} = (\bar{z})(\bar{w})$.
(3) $\bar{\bar{z}} = z$.
(4) $|\bar{z}| = |z|$.
(5) $\bar{z} = z$ if and only if z is real.

Properties (1) through (4) are computations. For (5) suppose $z = x + \mathbf{i}y$. Then $\bar{z} = x - \mathbf{i}y$, and $z = \bar{z}$ if and only if $y = -y$. But this is possible if and only if $y = 0$, and then $z = x + 0\mathbf{i}$ and z is real.

Now consider
$$z\bar{z} = (x + \mathbf{i}y)(x - \mathbf{i}y) = x^2 - \mathbf{i}^2 y^2 = x^2 + y^2 = |z|^2.$$
Therefore we have the following result.

Lemma 2.2.3
If $z \in \mathbb{C}$, then $z\bar{z} = |z|^2$.

From this lemma we can see that if $z \neq 0$, then
$$z \frac{\bar{z}}{|z|^2} = \frac{|z|^2}{|z|^2} = 1,$$
and so we make the following definition.

Definition 2.2.2
$\frac{1}{z} = \frac{\bar{z}}{|z|^2}$ for any $z \in \mathbb{C}$ with $z \neq 0$.

EXAMPLE 2.2.3 □

Let $z = 3 + 4\mathbf{i}$ and $w = 7 - 2\mathbf{i}$. Then
(1) $\frac{1}{z} = \frac{3-4\mathbf{i}}{3^2+4^2} = \frac{3}{25} - \frac{4}{25}\mathbf{i}$.
(2) $\frac{z}{w} = z\frac{1}{w} = (3+4\mathbf{i})\frac{7+2\mathbf{i}}{53} = \frac{13+34\mathbf{i}}{53} = \frac{13}{53} + \frac{34}{53}\mathbf{i}$.

From the definition we now have division of complex numbers and thus $\mathbb{C}$ is a field.

Theorem 2.2.1
$\mathbb{C}$ is a field. $\mathbb{R}$ is a subfield of $\mathbb{C}$, and the degree of $\mathbb{C}$ as an extension field of $\mathbb{R}$ is 2. Further, $\mathbb{C}$ is not an ordered field.

The degree being 2 comes from the easily observed fact that as a vector space over $\mathbb{R}$, $\{1, \mathbf{i}\}$ forms a basis. $\mathbb{C}$ cannot be an ordered field for the following reason. In any ordered field F, $1^2 = 1$ so 1 is positive. But then, from the trichotomy law -1 must be negative and therefore, as explained in the last section, cannot have a squareroot.

2.3 Geometrical Representation of Complex Numbers

To each complex number $z = x + \mathbf{i}y$ we can identify the point (x, y) in the xy-plane $\mathbb{R}^2$. Conversely, to each point $(x, y) \in \mathbb{R}^2$ we can identify the complex number $z = x + \mathbf{i}y$. When thought of in this way, as consisting of complex numbers, $\mathbb{R}^2$ is called the **complex plane**.

2.3. Geometrical Representation of Complex Numbers

Alternatively, we can think of the complex number $z = x + \mathbf{i}y$ as the vector $\vec{v} = (x, y)$, that is, the vector with representative starting at $(0,0)$ and ending at (x, y). In this interpretation $|\ |$ is just the magnitude of the vector (x, y), which is just the distance from the point (x, y) to the origin. The conjugate $\bar{z} = x - \mathbf{i}y$ is just the point $(x, -y)$, which is the point (x, y) reflected through the x-axis.

EXAMPLE 2.3.1
(see figure 2.1)

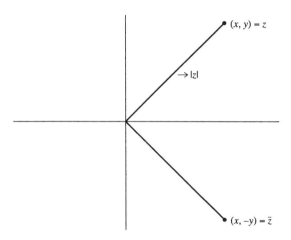

Figure 2.1. Geometrical Representation

We can describe the arithmetic operations in terms of this geometrical interpretation. Since addition and subtraction are done componentwise, addition and subtraction of complex numbers corresponds to the same vector operations as pictured in the diagram below.

EXAMPLE 2.3.2

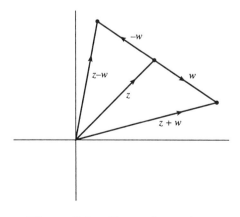

Figure 2.2. Vector Operations

Multiplication by real numbers is scalar multiplication of two-dimensional vectors. Geometrically this is a **stretching.** That is, if $z \in \mathbb{C}, r \in \mathbb{R}$ then:

(1) if $r \geq 0$, $rz = w$, where w is the vector in the same direction as z with magnitude $|r||z|$.
(2) if $r < 0$, $rz = w$ where w is the vector in the opposite direction as z with magnitude $|r||z|$.

If $z = x + \mathbf{i}y$, then $\mathbf{i}z = -y + \mathbf{i}x$. That is, multiplication by $\mathbf{i}$ takes the point (x, y) to the point $(-y, x)$. Since the inner product $< (x, y), (-y, x) > = -xy + xy = 0$, these vectors are orthogonal. Since $1\mathbf{i} = \mathbf{i}$, multiplication by $\mathbf{i}$ corresponds to a counterclockwise rotation by $90°$. Therefore, $\mathbf{i}$ is not really "imaginary" in any sense, it corresponds to a rotation.

Putting all this together we can give a complete geometric interpretation to complex multiplication. Suppose $z, w \in \mathbb{C}$ with $z = x + \mathbf{i}y$. Then consider $zw = (x + \mathbf{i}y)w = xw + \mathbf{i}yw$. Geometrically then, we first stretch the vector w by x, then stretch the vector w by y and rotate the second stretched vector by $90°$ counterclockwise. Finally we add the resulting vectors (see Figure 2.3).

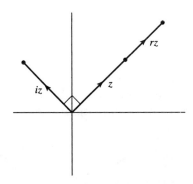

Figure 2.3. Complex Multiplication

2.4 Polar Form and Euler's Identity

If $P \in \mathbb{R}^2$ with rectangular coordinates (x, y), then P also has **polar coordinates** (r, θ) where r is the distance from the origin O to P and θ is the angle the vector $\overrightarrow{OP}$ makes with the positive x-axis (see Figure 2.4).

2.4. Polar Form and Euler's Identity

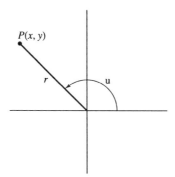

Figure 2.4. Polar Form

Here we restrict θ to be in the range $[0, 2\pi)$ so that each point in $\mathbb{R}^2$ has only one set of polar coordinates. The rectangular coordinates (x, y) of a point P are related to its polar coordinates (r, θ) by the formulas:

(i) $x = r \cos \theta$.
(ii) $y = r \sin \theta$.
(iii) $r = \sqrt{x^2 + y^2}$.   (2.4.1)
(iv) $\theta = \arctan(y/x)$ chosen in the appropriate quadrant.

If $z = x + \mathbf{i} y$ corresponds to the point (x, y) with polar coordinates (r, θ), then from (2.4.1), z can be written as

$$z = r(\cos \theta + \mathbf{i} \sin \theta) \quad (2.4.2.)$$

This is called the **polar form** for z. The angle θ is the **argument of z**, denoted by Arg z, and in this context $|z|$ is called the **modulus of z.**

EXAMPLE 2.4.1
Suppose $z = 1 - \mathbf{i}$; then $|z| = \sqrt{2}$ and Arg $z = \arctan(-1) = 7\pi/4$ since $(1,-1)$ is in the fourth quadrant. Therefore, $z = \sqrt{2}(\cos(7\pi/4) + \mathbf{i} \sin(7\pi/4))$. □

There is a very nice exponential way to express the polar form due to Euler. Before we describe this, we must look more closely at the powers of $\mathbf{i}$.

Now, $\mathbf{i}^2 = -1$ so $\mathbf{i}^3 = \mathbf{i}^2 \mathbf{i} = -\mathbf{i}$. Then $\mathbf{i}^4 = \mathbf{i}^3 \mathbf{i} = -\mathbf{i}^2 = 1$ and therefore $\mathbf{i}^5 = \mathbf{i}$. From this it follows that the powers of $\mathbf{i}$ repeat cyclically as $\{1, \mathbf{i}, -1, -\mathbf{i}\}$, and $\mathbf{i}^m = \mathbf{i}^n$ if and only if $n \equiv m$ (modulo 4). For example, $\mathbf{i}^{51} = \mathbf{i}^3 = -\mathbf{i}$. Further, the multiplicative inverse of any power of $\mathbf{i}$ is another power of $\mathbf{i}$, and so these powers form a group under multiplication.

Lemma 2.4.1
The powers of $\mathbf{i}$ form a cyclic group of order 4 under multiplication.

This result is actually a special case of a result about primitive nth roots of unity that we will discuss in the next section.

If t is a variable, recall that the functions e^t, $\sin t$, $\cos t$ have the following power series expansions.

(1) $e^t = 1 + t + t^2/2! + \cdots + t^n/n! + \cdots$.
(2) $\sin t = t - t^3/3! + t^5/5! - \cdots + (-1)^n t^{2n+1}/(2n+1)! + \cdots$. (2.4.3)
(3) $\cos t = 1 - t^2/2! + t^4/4! - \cdots + (-1)^n t^{2n}/(2n)! + \cdots$.

Now consider $t = \mathbf{i}\,\theta$ with θ real, and substitute into (2.4.3) to find $e^{\mathbf{i}\theta}$. (Although t is not a real variable, we do this formally.)

$$e^{\mathbf{i}\theta} = 1 + (\mathbf{i}\,\theta) + (\mathbf{i}\,\theta)^2/2! + \ldots = 1 + (\mathbf{i}\,\theta) - \theta^2/2! - \mathbf{i}\,\theta^3/3! + \cdots.$$

using the rules for the powers of $\mathbf{i}$. Then

$$e^{\mathbf{i}\theta} = (1 - \theta^2/2! + \theta^4/4! + \cdots) + \mathbf{i}(\theta - \theta^3/3! + \theta^5/5! + \cdots)$$
$$= \cos(\theta) + \mathbf{i}\sin(\theta).$$

This is known as **Euler's identity.**

Euler's Identity
$e^{\mathbf{i}\theta} = \cos(\theta) + \mathbf{i}\sin(\theta)$ for $\theta \in \mathbb{R}$.

Now, if $r = |z|$ and $\theta = \text{Arg } z$, we then have

$$z = r(\cos\theta + \mathbf{i}\sin\theta) = re^{\mathbf{i}\theta}. \tag{2.4.4}$$

This last identity makes multiplication of complex numbers very simple. Suppose $z = r_1 e^{\mathbf{i}\theta_1}$, $w = r_2 e^{\mathbf{i}\theta_2}$. Then $zw = r_1 r_2 e^{\mathbf{i}(\theta_1 + \theta_2)}$. Breaking this into components, we then have $|zw| = |z||w|$ and $\text{Arg}(zw) = \text{Arg } z + \text{Arg } w$.

Lemma 2.4.2
If $z, w \in \mathbb{C}$, then $|zw| = |z||w|$ and $\text{Arg}(zw) = \text{Arg } z + \text{Arg } w$.

Notice that $\text{Arg } \mathbf{i} = \pi/2$, and multiplication $\mathbf{i}\,z$ rotates z by 90°. That is, $\text{Arg}(\mathbf{i}\,z) = \pi/2 + \text{Arg } z = \text{Arg } \mathbf{i} + \text{Arg } z$, which follows directly from the lemma.

Euler's identity leads directly to what is called **Euler's magic formula.** Suppose $\theta = \pi$. Then

$$e^{\mathbf{i}\pi} = \cos(\pi) + \mathbf{i}\sin(\pi) = -1 + 0\,\mathbf{i} = -1.$$

Put succinctly, $e^{\mathbf{i}\pi} + 1 = 0$.

Euler's Magic Formula. $e^{\mathbf{i}\pi} + 1 = 0$.

This has been called a "magic" formula because the five most important numbers in mathematics – 0, 1, e, $\mathbf{i}$, π – are tied together in a very simple equation. If one thinks about how diversely these five numbers appear – 0 as the additive identity, 1 as the multiplicative identity, e as the natural exponential base, π as the ratio of the circumference to the diameter of any circle, and $\mathbf{i}$ as the imaginary unit, this result is truly amazing.

2.5 DeMoivre's Theorem for Powers and Roots

If $z = re^{\mathbf{i}\theta} \in \mathbb{C}$ and $n \in \mathbb{N}$, then $z^n = r^n e^{\mathbf{i} n\theta}$. Notice then that $|z^n| = |z|^n$ and $\mathrm{Arg}(z^n) = n\,\mathrm{Arg}\,z$. This is known as **DeMoivre's theorem for powers**.

DeMoivre's Theorem for Powers
If $z = re^{\mathbf{i}\theta} \in \mathbb{C}$ and $n \in \mathbb{N}$, then $z^n = r^n e^{\mathbf{i} n\theta}$. In particular, $|z^n| = |z|^n$ and $\mathrm{Arg}(z^n) = n\,\mathrm{Arg}\,z$.

EXAMPLE 2.5.1
Let $z = 1 - \mathbf{i}$ and let us find z^{10}.
Now, $|z| = \sqrt{2}$ and $\mathrm{Arg}\,z = 7\pi/4$, so $z = \sqrt{2}e^{\mathbf{i}\,7\pi/4}$. Therefore, $z^{10} = (\sqrt{2})^{10} e^{\mathbf{i}\,70\pi/4}$. Now, $(\sqrt{2})^{10} = 32$, while $70\pi/4 = 16\pi + 3\pi/2 \equiv 3\pi/2$ (as angles). It follows than that $z^{10} = 32 e^{\mathbf{i}\,3\pi/2} = 32(\cos(3\pi/2) + \mathbf{i}\sin(3\pi/2)) = -32\,\mathbf{i}$. □

Notice that since $\mathrm{Arg}(z^n) = n\,\mathrm{Arg}\,z$, the function $f(z) = z^n$ *winds* the complex number z around the origin. In Chapter 8 we give a proof of the Fundamental Theorem of Algebra based on this observation.

If $n \in \mathbb{N}, z \in \mathbb{C}$ and $w^n = z$ then w is an **nth root of z** which we denote by $z^{1/n}$. If $r \in \mathbb{R}, r > 0$ then there is exactly one real nth root of r. If $z = re^{\mathbf{i}\theta}$, then $w = r^{1/n} e^{\mathbf{i}\theta/n}$ satisfies $w^n = z$ and is one nth root of the complex number z. However, if $\frac{\theta + 2\pi k}{n} < 2\pi$ for $k \in \mathbb{N}$, then $w_k = r^{1/n} e^{\mathbf{i}\,\frac{\theta + 2\pi k}{n}}$ is also an nth root of z and is different from w. There are precisely n values of k that produce different nth roots of z for example $k = 0, 1, \ldots, n-1$. The n different nth roots of z are then

$$w_k = r^{1/n} e^{\mathbf{i}\,\frac{\theta + 2\pi k}{n}}, k = 0, 1, \ldots, n-1.$$

We have therefore proved the following theorem, which is known as **DeMoivre's theorem for roots**.

DeMoivre's Theorem for Roots
If $z \in \mathbb{C}, z \neq 0$, then there are exactly n distinct nth roots of z. If $z = re^{\mathbf{i}\theta}$, these are given by

$$w_k = r^{1/n} e^{\mathbf{i}\,\frac{\theta + 2\pi k}{n}}, k = 0, 1, \ldots, n-1.$$

EXAMPLE 2.5.2
Let $z = 1 + i$. Find the sixth roots of z.

Now $|z| = \sqrt{2}$, Arg z = arctan $1 = \pi/4$ so $z = \sqrt{2}e^{i\pi/4}$. The sixth roots of z are then

(1) $w_1 = 2^{1/12}e^{i\frac{\pi/4}{6}} = 2^{1/12}(\cos(\pi/24) + i\sin(\pi/24))$.
(2) $w_2 = 2^{1/12}e^{i\frac{\pi/4+2\pi}{6}} = 2^{1/12}(\cos(9\pi/24) + i\sin(9\pi/24))$.
(3) $w_3 = 2^{1/12}e^{i\frac{\pi/4+4\pi}{6}} = 2^{1/12}(\cos(17\pi/24) + i\sin(17\pi/24))$.
(4) $w_4 = 2^{1/12}e^{i\frac{\pi/4+6\pi}{6}} = 2^{1/12}(\cos(25\pi/24) + i\sin(25\pi/24))$.
(5) $w_5 = 2^{1/12}e^{i\frac{\pi/4+8\pi}{6}} = 2^{1/12}(\cos(33\pi/24) + i\sin(33\pi/24))$.
(6) $w_6 = 2^{1/12}e^{i\frac{\pi/4+10\pi}{6}} = 2^{1/12}(\cos(41\pi/24) + i\sin(41\pi/24))$. □

If $z = 1$, then an nth root, w, is called an **nth root of unity**. Since $|z| = 1$, then $|w| = 1$ for any nth root of unity, and therefore the roots of unity will differ only in their angles.

Corollary 2.5.1
There are exactly n distinct nth roots of unity given by

$$w_k = e^{i\frac{2\pi k}{n}} = \cos\left(\frac{2\pi k}{n}\right) + i\sin\left(\frac{2\pi k}{n}\right), \quad k = 0, 1, \ldots, n-1.$$

Geometrically then, nth roots of unity all fall on a circle of radius 1 and are located at the vertices of an inscribed regular n-gon with one vertex on the positive real axis. In the Figure 2.5 we picture the six, sixth roots of unity forming the vertices of a regular hexagon. For $z \neq 0$ with $|z| \neq 1$ the nth roots are at the vertices of a regular n-gon on a circle of radius $|z|^{1/n}$.

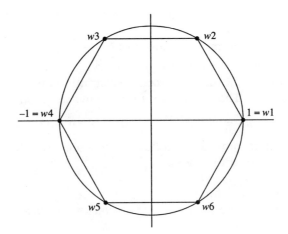

Figure 2.5. Sixth Roots of Unity

Let $w = e^{i\,2\pi/n}$, the nth root of unity with the smallest positive angle. This is called the **principal primitive nth root of unity.** The other nth roots are the distinct powers of w, that is, $\{w, w^2, w^3, \ldots, w^n = 1\}$. Further these nth roots of unity form a group under multiplication. (Notice that w is not the only primitive nth root of unity – see the exercises.)

Corollary 2.5.2
The nth roots of unity form a cyclic group of order n under multiplication. They are generated by w, the nth root of unity with the smallest positive angle.

The result we looked at before concerning the powers of $\mathbf{i}$ is just a special case of this corollary since $\mathbf{i}$ is the principal primitive fourth root of unity.

Exercises

2.1. Verify that $\mathbb{Q}(\sqrt{2})$ is a field.

2.2. Solve the equation $3x + 7 = 6$ in $\mathbb{Z}_{13}$. (Hint: Find the value of $1/3$ in $\mathbb{Z}_{13}$).

2.3. Prove that $\mathbb{Z}_n$ is a field if and only if n is a prime.

2.4. If $d \in \mathbb{Q}$ and d is not a perfect square show that $\mathbb{Q}(\sqrt{d})$ is a subfield of $\mathbb{R}$. (If d is a perfect square then $\mathbb{Q}(\sqrt{d}) = \mathbb{Q}$.)

2.5. Let $w^3 = 1$, $w \neq 1$. Show then that $1+w+w^2 = 0$. (Hint: Factor $w^3 - 1 = 0$.)

2.6. Let $w^3 = 1$, $w \neq 1$, so that from Exercise 2.5, $1+w+w^2 = 0$, or $w^2 = -1 - w$. Let $\mathbb{Q}(w) = \{a + bw; a, b \in \mathbb{Q}\}$. Define arithmetic on $\mathbb{Q}(w)$ by algebraic manipulation. Show that $\mathbb{Q}(w)$ is a subfield of $\mathbb{C}$. (Hint: Use $w^2 = -1 - w$.) What is the degree of $\mathbb{Q}(w)$ over $\mathbb{Q}$?

2.7. This exercise is to show that there are infinitely many transcendentals.
   (i) For $n \in \mathbb{N}$ let $\mathcal{P}_n$ =[all rational polynomials of degree $\leq n$]. Show that $\mathcal{P}_n$ is countable. (Hint: $\mathcal{P}_n$ can be considered as all $(n+1)$-tuples with entries from $\mathbb{Q}$. Recall that if A, B are countable then the Cartesian product $A \times B$ is also countable.)
   (ii) For $n \in \mathbb{N}$ let $\mathcal{Q}_n$ =[all roots for polynomials in $\mathcal{P}_n$]. Show that $\mathcal{Q}_n$ is countable. (Hint:Use the facts that a polynomial of degree n has at most n roots and countable unions of countable sets are countable.)
   (iii) Let $\mathcal{A}$ be the set of algebraic numbers over $\mathbb{Q}$. Show that $\mathcal{A}$ is countable. (Hint: $\mathcal{A} = \bigcup_1^\infty \mathcal{Q}_n$.)
   (iv) Let $\mathcal{T}$ be the set of transcendental numbers over $\mathbb{Q}$. Show that $\mathcal{T}$ is uncountable. (Hint: $\mathbb{R} = \mathcal{A} \cup \mathcal{T}$. Suppose $\mathcal{T}$ were countable.)

2.8. Prove that the LUB property for $\mathbb{R}$ is equivalent to the nested intervals property.

2.9. Let $z = 4 + 7\mathbf{i}$, $w = 6 - \mathbf{i}$. Find

(1) $\bar{z}, \bar{w}, |z|, |w|$.
(ii) $z + w, zw, z/w$.
(iii) $\sqrt{z}$ (2 solutions).
(iv) z in polar form.
(v) z^5.
(vi) Solve the equation for Z: $zZ + 6w = 1 - i$..

2.10. Prove Lemma 2.2.1.

2.11. Prove Lemma 2.2.2.

2.12. In Exercise 2.9 multiply $zw = 4w + \mathbf{i}\,7w$ and trace out the geometrical steps on w.

2.13. In Exercise 2.9 multiply z and w by putting them in polar form. Estimate Arg z and Arg w by calculator.

2.14. Let $z = 1 + \sqrt{3}\,\mathbf{i}$. Find:
(i) z^{26}
(ii) The five distinct fifth roots of z
(iii) Plot the answers to (ii) geometrically.

2.15. Let $1, w, w^2, \ldots, w^{n-1}$ form a cyclic group G of order n. Show that w^k also generates G if and only if $(k, n) = 1$. (Hint: If $(k, n) = 1$ that is k and n are relatively prime then 1 can be written as a linear combination of k and n.)

2.16. Let $w = e^{2\pi/8}$ be the principal primitive eighth root of unity. What powers of w also generate the group of eighth roots?

3 Polynomials and Complex Polynomials

CHAPTER

3.1 The Ring of Polynomials over a Field

If F is a field and n is a nonnegative integer, then a **polynomial of degree n over F** is a formal sum of the form

$$P(x) = a_0 + a_1 x + \cdots + a_n x^n \tag{3.1}$$

with $a_i \in F$ for $i = 0, .., n$, $a_n \neq 0$, and x an indeterminate. A **polynomial** $P(x)$ over F is either a polynomial of some degree or the expression $P(x) = 0$, which is called the **zero polynomial** and has no degree. We denote the degree of $P(x)$ by **deg $P(x)$**. A polynomial of zero degree has the form $P(x) = a_0$ and is called a **constant polynomial** and can be identified with the corresponding element of F. The elements $a_i \in F$ are called the **coefficients of** $P(x)$; a_n is the **leading coefficient**. If $a_n = 1$, $P(x)$ is called a **monic polynomial**. Two nonzero polynomials are equal if and only if they have the same degree and exactly the same coefficients. A polynomial of degree 1 is called a **linear polynomial** while one of degree two is a **quadratic polynomial**.

We denote by $F[x]$ the set of all polynomials over F and by $F_n[x]$ the set of polynomials over F of degree less than or equal to n together with the zero polynomial.

$$F[x] = \{P(x); P(x) \text{ is a polynomial over } F\}. \tag{3.2}$$

$$F_n[x] = \{P(x) \in F[x]; deg P(x) \leq n \text{ or } P(x) = 0\}. \tag{3.3}$$

We will see that $F[x]$ becomes a ring with much the same properties as the integers $\mathbb{Z}$. We first define addition, subtraction, and multiplication on $F[x]$ by algebraic manipulation. That is, suppose $P(x) = a_0 + a_1 x + \cdots + a_n x^n$, $Q(x) = b_0 + b_1 x + \cdots + b_m x^m$ then

$$P(x) \pm Q(x) = (a_0 \pm b_0) + (a_1 \pm b_1)x + \cdots \qquad (3.4)$$

that is, the coefficient of x^i in $P(x) \pm Q(x)$ is $a_i \pm b_i$.

$$P(x)Q(x) = (a_0 b_0) + (a_1 b_0 + a_0 b_1)x + (a_0 b_2 + a_1 b_1 + a_2 b_0)x^2 + \cdots + (a_n b_m)x^{n+m} \qquad (3.5)$$

that is, the coefficient of x^i in $P(x)Q(x)$ is $(a_0 b_i + a_1 b_{i-1} + \cdots + a_i b_0)$.

EXAMPLE 3.1.1
Let $P(x) = 3x^2 + 4x - 6$ and $Q(x) = 2x + 7$ be in $\mathbb{Q}[x]$. Then

$$P(x) + Q(x) = 3x^2 + 6x + 1$$

and

$$P(x)Q(x) = (3x^2 + 4x - 6)(2x + 7) = 6x^3 + 29x^2 + 16x - 42. \qquad \square$$

From the definitions the following degree relationships are clear. The proofs are in the exercises.

Lemma 3.1.1
Let $P(x) \neq 0, Q(x) \neq 0 \in F[x]$. Then:

(1) $\deg P(x)Q(x) = \deg P(x) + \deg Q(x)$.
(2) $\deg (P(x) \pm Q(x)) \leq \text{Max}(\deg P(x), \deg Q(x))$.

In a ring R a **zero divisor** is a nonzero element $r \in R$ for which there exists another *nonzero* element $s \in R$ with $rs = 0$. For example, in the ring of integers modulo 6, $\mathbb{Z}_6$, the elements 2, 3 are both nonzero but $(2)(3) = 0$ and hence both are zero divisors. A commutative ring with an identity, having *no zero-divisors*, is an **integral domain**. Examples of integral domains are $\mathbb{Z}, \mathbb{Q}, \mathbb{R}, \mathbb{C}$, and $\mathbb{Z}_p$ if p is a prime. In general, a field must be an integral domain, as we show in the next lemma. Recall that an element in a ring with a multiplicative inverse is called a **unit**.

Lemma 3.1.2
If R is a commutative ring with an identity and $r \in R$ is a unit, then r is not a zero divisor. In particular, if R is a field, then R is an integral domain.

Proof
Suppose $r \in R$ is a unit and $rs = 0$. Since r is unit there exists r^{-1} such that $r^{-1} r = 1$. Then $r^{-1}(rs) = r^{-1} 0 = 0$ while $r^{-1}(rs) = (r^{-1} r)s = 1s = s$. Hence $s = 0$, and r is not a zero divisor.

3.1. The Ring of Polynomials over a Field

A field F is a commutative ring with an identity in which every nonzero element is a unit. From the above, no nonzero element can be a zero divisor, and hence F is an integral domain. ∎

We now obtain the following.

Theorem 3.1.1
If F is a field, then $F[x]$ forms a commutative ring with identity (actually an integral domain). F can be naturally embedded into $F[x]$ by identifying each element of F with the corresponding constant polynomial. The only units in $F[x]$ are the nonzero elements of F. $F_n[x]$ forms a vector space of dimension $n+1$ over F.

Proof
To verify the basic ring properties is solely computation and is left to the exercises. Since $\deg P(x)Q(x) = \deg P(x) + \deg Q(x)$, it follows that if neither $P(x) \neq 0$ nor $Q(x) \neq 0$ then $P(x)Q(x) \neq 0$ and therefore $F[x]$ is an integral domain.

If $G(x)$ is a unit in $F[x]$, then there exists an $H(x) \in F[x]$ with $G(x)H(x) = 1$. From the degrees we have $\deg G(x) + \deg H(x) = 0$ and since $\deg G(x) \geq 0, \deg H(x) \geq 0$. This is possible only if $\deg G(x) = \deg H(x) = 0$. Therefore $G(x) \in F$.

Finally, from Lemma 3.1.1, $F_n[x]$ is closed under addition and subtraction and thus is an abelian group. Multiplication by elements of F doesn't raise the degree, so $F_n[x]$ admits scalar multiplication from F and therefore forms a vector space over F. The set $\{1, x, x^2, \ldots, x^n\}$ constitutes a basis, so $dim(F_n[x]) = n+1$. ∎

A polynomial $P(x) \in F[x]$ can also be considered as a function $P: F \to F$ via the **substitution process**. If $P(x) = a_0 + a_1 x + \cdots + a_n x^n \in F[x]$ and $t \in F$, then

$$P(t) = a_0 + a_1 t + \cdots + a_n t^n \in F$$

since F is closed under all the operations used in the polynomial. If the field F is closed in some sense – such as the Cauchy sequence property – this function is continuous. In particular, real and complex polynomials are continuous functions, $\mathbb{R} \to \mathbb{R}$, $\mathbb{C} \to \mathbb{C}$, respectively.

If $r \in F, P(x) \in F[x]$, and $P(r) = 0$ under the substitution process, we say that r is a **root** of $P(x)$ or **zero** of $P(x)$. Synonymously, we say that, *r satisfies $P(x)$*. As pointed out in the introduction, many authors prefer to use the word *root* for equations and the word *zero* for polynomials and functions. Throughout this book we will use the word *root*. Historically, this has been most often connected with the Fundamental Theorem of Algebra. In the next section we will see that r being a root implies a certain factorability property.

3.2 Divisibility and Unique Factorization of Polynomials

For a field F, the algebraic structure of the polynomial ring $F[x]$ is surprisingly similar to that of the integers $\mathbb{Z}$. Recall that $\mathbb{Z}$ satisfies the following basic property, called the **fundamental theorem of arithmetic**.

Fundamental Theorem of Arithmetic
If $z \in \mathbb{Z}$, then z admits a factorization into primes. Further, this factorization is unique up to ordering and unit factors.

In general, an integral domain R that admits unique factorization (unique up to ordering and unit factors) into primes is called a **unique factorization domain** abbreviated **UFD**. The fundamental theorem of arithmetic then says that $\mathbb{Z}$ is a UFD. We will now show that for any field F, the polynomial ring $F[x]$ is a UFD. First we must define factorization and primes.

Definition 3.2.1
If $f(x), g(x) \in F[x]$ with $g(x) \neq 0$ then $g(x)$ **divides** $f(x)$, or $g(x)$ is a **factor** of $f(x)$, if there exists a polynomial $q(x) \in F[x]$ such that $f(x) = q(x)g(x)$. We denote this by $g(x)|f(x)$.

If $0 \neq f(x)$ has no nontrivial, nonunit factors (it cannot be factorized into polynomials of lower degree); then $f(x)$ is an **irreducible polynomial**, or **prime polynomial**. Clearly, if $\deg g(x) = 1$ then $g(x)$ is irreducible.

The fact that $F[x]$ is a UFD follows from the division algorithm for polynomials, which is entirely analogous to the division algorithm for integers.

Division Algorithm in $F[x]$
If $0 \neq f(x), 0 \neq g(x) \in F[x]$ then there exist unique polynomials $q(x), r(x) \in F[x]$ such that $f(x) = q(x)g(x) + r(x)$ where $r(x) = 0$ or $\deg r(x) < \deg g(x)$. (The polynomials $q(x)$ and $r(x)$ are called respectively the quotient and remainder.)

This theorem is essentially long division of polynomials. A formal proof is based on induction on the degree of $g(x)$. We omit this but give some examples from $\mathbb{Q}[x]$.

3.2. Divisibility and Unique Factorization of Polynomials

EXAMPLE 3.2.1

(a) Let $f(x) = 3x^4 - 6x^2 + 8x - 6, g(x) = 2x^2 + 4$. Then

$$\frac{3x^4 - 6x^2 + 8x - 6}{2x^2 + 4} = \frac{3}{2}x^2 - 6 \text{ with remainder } 8x + 18.$$

Thus, here $q(x) = \frac{3}{2}x^2 - 6, r(x) = 8x + 18$.

(b) Let $f(x) = 2x^5 + 2x^4 + 6x^3 + 10x^2 + 4x, g(x) = x^2 + x$. Then

$$\frac{2x^5 + 2x^4 + 6x^3 + 10x^2 + 4x}{x^2 + x} = 2x^3 + 6x + 4.$$

Thus here $q(x) = 2x^3 + 6x + 4$ and $r(x) = 0$. □

Using the division algorithm, the development of unique factorization follows as in $\mathbb{Z}$. We need the idea of a **greatest common divisor**, or **gcd**, and the following lemmas.

Definition 3.2.2

(1) If $f(x), g(x) \in F[x]$, then $d(x) \in F[x]$ is the **greatest common divisor**, or **gcd**, of $f(x), g(x)$ if $d(x)$ is monic, $d(x)$ divides both $g(x)$ and $f(x)$, and if $d_1(x)$ divides both $g(x)$ and $f(x)$ then $d_1(x)$ divides $d(x)$. We write $d(x) = (g(x), f(x))$. If $(f(x), g(x)) = 1$, then we say that $f(x)$ and $g(x)$ are **relatively prime**.

(2) An expression of the form $f(x)h(x) + g(x)k(x)$ is called a **linear combination** of $f(x), g(x)$.

Lemma 3.2.1
Given $f(x), g(x) \in F[x]$, then the gcd exists, is unique, and equals the monic polynomial of least degree that is expressible as a linear combination of $f(x), g(x)$.

Finding the gcd of two polynomials is done in the same manner as finding the gcd of two integers. That is, we use the **Euclidean algorithm**. This is done in the following manner. Suppose $0 \neq f(x), 0 \neq g(x) \in F[x]$. Use repeated applications of the division algorithm to obtain the sequence:

$$f(x) = q(x)g(x) + r(x),$$
$$g(x) = q_1(x)r(x) + r_1(x),$$
$$r(x) = q_2(x)r_1(x) + r_2(x),$$
$$\ldots$$
$$\ldots$$
$$r_{k-1}(x) = q_{k+1}(x)r_k(x).$$

Since each division reduces the degree, and the degree is finite, this process will ultimately end. Let $r_k(x)$ be the last nonzero remainder polynomial and suppose c is the leading coefficient of $r_k(x)$. Then $c^{-1}r_k(x)$ is the gcd. We give an example.

EXAMPLE 3.2.2
In $\mathbb{Q}[x]$ find the gcd of the polynomials $f(x) = x^3 - 1$ and $g(x) = x^2 - 2x + 1$ and express it as a linear combination of the two. □

Using the Euclidean algorithm we obtain
$$x^3 - 1 = (x^2 - 2x + 1)(x + 2) + (3x - 3),$$
$$x^2 - 2x + 1 = (3x - 3)\left(\frac{1}{3}x - \frac{1}{3}\right).$$

Therefore, the last nonzero remainder is $3x - 3$, and hence the gcd is $x - 1$. Working backwards we have
$$3x - 3 = (x^3 - 1) - (x^2 - 2x + 1)(x + 2)$$
so
$$x - 1 = \frac{1}{3}(x^3 - 1) - \frac{1}{3}(x^2 - 2x + 1)(x + 2)$$
expressing the gcd as a linear combination of the two given polynomials.

Lemma 3.2.2
(Euclid's Lemma) If $p(x)$ is an irreducible polynomial and $p(x)$ divides $f(x)g(x)$, then $p(x)$ divides $f(x)$ or $p(x)$ divides $g(x)$.

Proof
Suppose $p(x)$ does not divide $f(x)$. Then since $p(x)$ is irreducible, $p(x)$ and $f(x)$ must be relatively prime. Therefore, there exist $h(x), k(x)$ such that
$$f(x)h(x) + p(x)k(x) = 1.$$
Multiply through by $g(x)$ to obtain
$$g(x)f(x)h(x) + g(x)p(x)k(x) = g(x).$$
Now, $p(x)$ divides each term on the left-hand side since $p(x)|g(x)f(x)$ and therefore $p(x)|g(x)$. ∎

EXAMPLE 3.2.3
Show that $x^2 + x + 1$ is irreducible over $\mathbb{Q}[x]$ and $\mathbb{R}[x]$ but not irreducible over $\mathbb{C}[x]$.
    Suppose $f(x) = x^2 + x + 1$ had a nontrivial factor in $\mathbb{R}[x]$. Since deg $f(x) = 2$, this factor must have degree 1 and $f(x)$ would factor into two

3.3. Roots of Polynomials and Factorization

linear factors

$$f(x) = x^2 + x + 1 = (ax + b)(cx + d).$$

Then $f(x)$ has a real root, namely $-b/a$. However, from the quadratic formula the roots are

$$\omega_1 = \frac{-1 + \sqrt{3}i}{2}, \omega_2 = \frac{-1 - \sqrt{3}i}{2}$$

which are both nonreal. Therefore, $f(x)$ must be irreducible over $\mathbb{R}[x]$ and also over $\mathbb{Q}[x]$.

A computation shows that in $\mathbb{C}[x]$, $f(x) = (x - \omega_1)(x - \omega_2)$ and therefore it factors in $\mathbb{C}[x]$. □

Theorem 3.2.1
If $0 \neq f(x) \in F[x]$, then $f(x)$ has a factorization into irreducible polynomials that is unique up to ordering and unit factors. In other words $F[x]$ is a UFD.

The proof is almost identical to the proof for $\mathbb{Z}$, and we sketch it. First we use induction on the degree of $f(x)$ to obtain a prime factorization. If $\deg f(x) = 1$, then $f(x)$ is irreducible, so suppose $\deg f(x) = n > 1$. If $f(x)$ is irreducible, then it has such a prime factorization. If $f(x)$ is not irreducible, then $f(x) = h(x)g(x)$ with $\deg g(x) < n$ and $\deg h(x) < n$. By the inductive hypothesis, both $g(x)$ and $h(x)$ have prime factorizations, and so $f(x)$ does as well.

Now suppose that $f(x)$ has two prime factorizations

$$f(x) = p_1(x)^{n_1} \ldots p_k(x)^{n_k} = q_1(x)^{m_1} \ldots q_t(x)^{m_t}$$

where $p_i(x), i = 1, \ldots, n$, $q_j(x), j = 1, \ldots, t$ are prime polynomials. Consider $p_i(x)$. Then $p_i(x) | q_1(x)^{m_1} \ldots q_t(x)^{m_t}$, and hence from Euclid's lemma, $p_i(x) | q_j(x)$ for some j. Since both are irreducible, $p_i(x) = cq_j(x)$ for some unit c. By repeated application of this argument we get that $n_i = m_j$. Thus we have the same primes with the same multiplicities but perhaps unit factors, proving the theorem.

This whole development could be done starting with just a ring R rather than a field F. It can be proved that if R is a UFD, $R[x]$ is also a UFD. Thus, for example, $\mathbb{Z}[x]$ is a UFD. The proof is different and somewhat more complicated than that for fields.

3.3 Roots of Polynomials and Factorization

We now show that if $f(x)$ has a root and $\deg f(x) > 1$ then $f(x)$ factors — that is, $f(x)$ is not irreducible.

Lemma 3.3.1
If c is a root of $P(x)$ then $(x-c)$ divides $P(x)$ - that is, $P(x) = (x-c)Q(x)$ with $\deg Q(x) = \deg P(x) - 1$.

Proof
Suppose $P(c) = 0$. Then from the division algorithm $P(x) = (x-c)Q(x) + r(x)$ where $r(x) = 0$ or $r(x) = f \in F$, since $\deg r(x) < \deg(x-c) = 1$. Therefore
$$P(x) = (x-c)Q(x) + f.$$
Substituting, we have $P(c) = 0 + f = 0$, and so $f = 0$. Hence $P(x) = (x-c)Q(x)$. ∎

Corollary 3.3.1
An irreducible polynomial of degree greater than one over a field F has no roots in F.

From this we obtain the following important theorem.

Theorem 3.3.1
A polynomial of degree n in $F[x]$ can have at most n distinct roots.

Proof
Suppose $P(x)$ has degree n and suppose $c_1, \ldots, c_n$ are n distinct roots. From repeated application of Lemma 3.3.1,
$$P(x) = k(x-c_1)\ldots(x-c_n)$$
where $k \in F$. Suppose c is any other root. Then $P(c) = 0 = k(c-c_1)\ldots(c-c_n)$. Since a field F has no zero divisors, one of these terms must be zero: $c - c_i = 0$ for some i, and hence $c = c_i$. ∎

Besides having a maximum of n roots (with n the degree) the roots of a polynomial are unique. Suppose $P(x)$ has degree n and roots $c_1, .., c_k$ with $k \leq n$. Then from the unique factorization in $F[x]$ we have
$$P(x) = (x-c_1)^{m_1} \ldots (x-c_k)^{m_k} Q_1(x) \ldots Q_t(x),$$
where $Q_i(x), i = 1, .., t$ are irreducible and of degree greater than 1. The exponents m_i are called the **multiplicities** of the roots c_i. Let c be a root. Then as above,
$$(c-c_1)^{m_1} \ldots (c-c_k)^{m_k} Q_1(c) \ldots Q_t(c) = 0.$$
Now $Q_i(c) \neq 0$ for $i = 1, .., t$ since $Q_i(x)$ are irreducible of degree > 1. Therefore, $(c - c_i) = 0$ for some i, and hence $c = c_i$.

Corollary 3.3.2
If $P(x) \in F[x]$ with deg $P(x) = 2$, then $P(x)$ is irreducible if and only if $P(x)$ has no root in F.

Proof
Suppose $P(x)$ is irreducible. Since deg $P(x) > 1$, $P(x)$ can have no root in F. Conversely suppose $P(x)$ has no root in F. If $P(x)$ were not irreducible then it would factor into two linear factors,
$$P(x) = (ax+b)(cx+d).$$
But then $P(x)$ would have the root $-b/a \in F$. ∎

3.4 Real and Complex Polynomials

We now consider the underlying field to be $\mathbb{R}$ or $\mathbb{C}$ and consider real and complex polynomials – that is, polynomials in $\mathbb{R}[x]$ and $\mathbb{C}[x]$ respectively. We first need the following important result.

Theorem 3.4.1
A real polynomial of odd degree has a real root.

Proof
Suppose $P(x) \in \mathbb{R}[x]$ with deg $P(x) = n = 2k+1$ and suppose the leading coefficient $a_n > 0$ (the proof is almost identical if $a_n < 0$). Then
$$P(x) = a_n x^n + \text{(lower terms) and } n \text{ is odd.}$$
Then

(1) $\lim_{x \to \infty} P(x) = \lim_{x \to \infty} a_n x^n = \infty$ since $a_n > 0$.
(2) $\lim_{x \to -\infty} P(x) = \lim_{x \to -\infty} a_n x^n = -\infty$ since $a_n > 0$ and n is odd.

From (1) $P(x)$ gets arbitrarily large positively so there exists an x_1 with $P(x_1) > 0$. Similarly, from (2) there exists an x_2 with $P(x_2) < 0$.

A real polynomial is a continuous real-valued function for all $x \in \mathbb{R}$. Since $P(x_1)P(x_2) < 0$, it follows from the intermediate value theorem that there exists an x_3, between x_1 and x_2, such that $P(x_3) = 0$. ∎

As an immediate consequence we have the following corollary:

Corollary 3.4.1
If $P(x) \in \mathbb{R}[x]$ is irreducible and nonlinear, then its degree is even. (We will see later that it must be 2.)

Now we consider complex polynomials.

Lemma 3.4.1
Every degree 2 complex polynomial has a root in $\mathbb{C}$.

Proof
This is just the quadratic formula. If $P(x) = ax^2 + bx + c$, then the roots formally are

$$x_1 = \frac{-b + \sqrt{b^2 - 4ac}}{2a}, x_2 = \frac{-b - \sqrt{b^2 - 4ac}}{2a}.$$

From DeMoivre's theorem every complex number has a square root; hence x_1, x_2 exist in $\mathbb{C}$. They of course may be the same, if $b^2 - 4ac = 0$. ∎

To go further we need the concept of the **conjugate of a polynomial** and some straightforward consequences of this idea.

Definition 3.4.1
If $P(x) = a_0 + \cdots + a_n x^n$ is a complex polynomial then its **conjugate** is the polynomial $\overline{P}(x) = \overline{a_0} + \cdots + \overline{a_n} x^n$. That is, the conjugate is the polynomial whose coefficients are the conjugates of those of $P(x)$.

Lemma 3.4.2
For any $P(x) \in \mathbb{C}[x]$

(1) $\overline{P(z)} = \overline{P}(\overline{z})$ *if $z \in \mathbb{C}$*
(2) $P(x)$ *is a real polynomial if and only if $P(x) = \overline{P}(x)$*
(3) *If $P(x)Q(x) = H(x)$ then $\overline{H}(x) = (\overline{P}(x))(\overline{Q}(x))$*

Proof

(1) Suppose $z \in \mathbb{C}$ and $P(z) = a_0 + \cdots + a_n z^n$. Then

$$\overline{P(z)} = \overline{a_0 + \cdots + a_n z^n} = \overline{a_0} + \overline{a_1 z} + \cdots + \overline{a_n z^n} = \overline{P}(\overline{z}).$$

(2) Suppose $P(x)$ is real then $a_i = \overline{a_i}$ for all its coefficients and hence $P(x) = \overline{P}(x)$. Conversely suppose $P(x) = \overline{P}(x)$. Then $a_i = \overline{a_i}$ for all its coefficients and hence $a_i \in \mathbb{R}$ for each a_i and so $P(x)$ is a real polynomial.
(3) The proof is a computation and left to the exercises. ∎

Lemma 3.4.3
Suppose $G(x) \in \mathbb{C}[x]$. Then $H(x) = G(x)\overline{G}(x) \in \mathbb{R}[x]$.

Proof
$\overline{H}(x) = \overline{G(x)\overline{G}(x)} = \overline{G}(x)\overline{\overline{G}}(x) = \overline{G}(x)G(x) = G(x)\overline{G}(x) = H(x)$. Therefore, $H(x)$ is a real polynomial. ∎

Lemma 3.4.4
If $f(x) \in \mathbb{R}[x]$ and $f(z_0) = 0$ then $f(\overline{z_0}) = 0$; the complex roots of real polynomials come in conjugate pairs.

Proof
$f(z_0) = 0$ implies that $\overline{f(z_0)} = 0$. But then $\overline{f}(\overline{z_0}) = 0$. Since $f(x)$ is real, $f(x) = \overline{f}(x)$, so $f(\overline{z_0}) = 0$. ■

Notice that if z_0 is a root of $f(x) \in \mathbb{R}[x]$ then both $(x - z_0)$ and $(x - \overline{z_0})$ divide $f(x)$.

Lemma 3.4.5
$(x - z)(x - \overline{z}) \in \mathbb{R}[x]$ for any $z \in \mathbb{C}$.

We leave the proof for the exercises. Notice that this now implies that any real polynomial of degree ≤ 3 completely factorizes over $\mathbb{C}$.

Finally, to complete this section, we prove the following theorem, which shows that to prove the Fundamental Theorem of Algebra we need only to prove that real polynomials must have complex roots.

Theorem 3.4.2
If every nonconstant real polynomial has at least one complex root, then every nonconstant complex polynomial has at least one complex root.

Proof
Let $P(x) \in \mathbb{C}[x]$ and suppose that every nonconstant real polynomial has at least one complex root. Let $H(x) = P(x)\overline{P}(x)$. From Lemma 3.4.3, $H(x) \in \mathbb{R}[x]$. By supposition there exists a $z_0 \in \mathbb{C}$ with $H(z_0) = 0$. Then $P(z_0)\overline{P}(z_0) = 0$, and since $\mathbb{C}$ has no zero divisors, either $P(z_0) = 0$ or $\overline{P}(z_0) = 0$. In the first case z_0 is a root of $P(x)$. In the second case $\overline{P}(z_0) = 0$, bu then from Lemma 3.4.2 $0 = \overline{P}(z_0) = \overline{\overline{P}(\overline{z_0})} = P(\overline{z_0})$. Therefore $\overline{z_0}$ is a root of $P(x)$. ■

Notice that this theorem is not the Fundamental Theorem of Algebra. It only says that to prove the Fundamental Theorem we need only prove it for real polynomials.

3.5 The Fundamental Theorem of Algebra - Proof One

We now present the first proof of the Fundamental Theorem of Algebra. This proof based solely on advanced calculus and from advanced calculus we need the next result.

3. Polynomials and Complex Polynomials

Lemma 3.5.1
If $f:D \to \mathbb{R}$ is continuous where D is a closed and bounded (compact) subset of $\mathbb{R}^2$, then $f(x,y)$ has a minimum and maximum value on D.

This is the two-dimensional version of the **extreme values theorem** from elementary calculus, which states that if $f : [a, b] \to \mathbb{R}$ is continuous then $f(x)$ has a minimum and a maximum on $[a, b]$. More generally this theorem holds for continuous functions $f : \mathbb{R}^n \to \mathbb{R}$ on compact domains for any $n \geq 1$. For a proof of Lemma 3.5.1, we refer the reader to any advanced calculus text.

Theorem 3.5.1
(The Fundamental Theorem of Algebra) If $f(x) \in \mathbb{C}[x]$, with $f(x)$ nonconstant, then $f(x)$ has at least one complex root.

Our proof depends on the next two lemmas.

Lemma 3.5.2
Let $f(x) \in \mathbb{C}[x]$. Then $|f(x)|$ takes on a minimum value at some point $z_0 \in \mathbb{C}$.

Proof
It is straightforward that as $|x| \to \infty$, $|f(x)| \to \infty$. Since $|f(x)|$ is large for large $|x|$ it follows that the greatest lower bound m of $|f(z)|$ for $z \in \mathbb{C}$ is also the greatest lower bound in some sufficiently large disk $|z| \leq r$.

Since $|f(x)|$ is a continuous real-valued function it follows from Lemma 3.5.1 that $|f(x)|$ will attain its minimum value on this disk. ∎

Lemma 3.5.3
Suppose $f(x) \in \mathbb{C}[x]$ with $f(x)$ nonconstant. If $f(x_0) \neq 0$, then $|f(x_0)|$ is not the minimum value of $|f(x)|$.

Proof
Let $f(x)$ be a nonconstant complex polynomial and suppose x_0 is a point with $f(x_0) \neq 0$. Make the change of variable $x + x_0$ for x. This shifts x_0 to the origin, so that we may assume that $f(0) \neq 0$. Next multiply $f(x)$ by $f(0)^{-1}$ so that $f(0) = 1$. We must then show that 1 is not the minimum value of $|f(x)|$.

Let k be the lowest nonzero power of x occurring in $f(x)$. Then $f(x)$ can be assumed to have the form

$$f(x) = 1 + ax^k + \text{ terms of degree } > k.$$

Now let α be a k-th root of $-a^{-1}$, which exists by DeMoivre's theorem. Make the final change of variable αx for x. Now $f(x)$ has the form

$$f(x) = 1 - x^k + x^{k+1}g(x) \text{ for some polynomial } g(x).$$

For small positive real x we obtain from the triangle inequality

$$|f(x)| \leq |1 - x^k| + x^{k+1}|g(x)|.$$

But $x^k < 1$ for small real x, so this has the form

$$|f(x)| \leq 1 - x^k + x^{k+1}|g(x)| = 1 - x^k(1 - x|g(x)|).$$

For small real x, $x|g(x)|$ is small, so then x_0 can be chosen so that $x_0|g(x_0)| < 1$. It follows that $x_0^k(1 - x_0|g(x_0)|) > 0$, so then $|f(x_0)| < 1 = |f(0)|$ completing the proof. ∎

Combining these two lemmas we obtain our first proof of the Fundamental Theorem of Algebra.

Proof
Let $f(x)$ be a nonconstant complex polynomial. From Lemma 3.5.2, $|f(x)|$ has a minimum value at some point $x_0 \in \mathbb{C}$. Then from Lemma 3.5.3 it follows that $|f(x_0)| = 0$, and hence $f(x_0) = 0$ for otherwise it would not be the minimum value. Therefore, $f(x)$ has a complex root. ∎

3.6 Some Consequences of the Fundamental Theorem

In the final section of this chapter we look at some consequences of the Fundamental Theorem.

Corollary 3.6.1
A complex polynomial completely factorizes into linear factors.

Proof
Let $f(x) \in \mathbb{C}[x]$ and use induction on the degree. The corollary is clearly true if $\deg f(x) = 1$, since then $f(x)$ is itself linear. Suppose $\deg f(x) = n$. From the Fundamental Theorem of Algebra, there exists a root x_0, and therefore $(x - x_0)$ divides $f(x)$. Hence $f(x) = (x - x_0)g(x)$ with $\deg g(x) < n$. From the inductive hypothesis, $g(x)$ factors into linear factors, so therefore $f(x)$ does also. ∎

Corollary 3.6.2
Suppose $f(x) \in \mathbb{C}[x]$ with $\deg f(x) = n$. Suppose the roots of $f(x)$ are $x_1, x_2, \ldots, x_n$ (some may be repeated). Then

$$f(x) = \alpha(x - x_1)\ldots(x - x_n), \alpha \in \mathbb{C}.$$

Corollary 3.6.3
A real polynomial factorizes into degree 1 and degree 2 factors. Equivalently, the only irreducible real polynomials are linear polynomials and quadratic polynomials without real roots.

Proof
Suppose $P(x) \in \mathbb{R}[x]$, then $P(x) \in \mathbb{C}[x]$. Suppose $z_1, \ldots, z_n$ are its complex roots, so that
$$P(x) = \alpha(x - z_1) \ldots (x - z_n)$$
where here $\alpha \in \mathbb{R}$, since α is the leading coefficient of $P(x)$.

If z_i is real, then $(x - z_i)$ is a real linear factor. If $z_i \notin \mathbb{R}$ then its complex conjugate $\overline{z_i}$ is also a root. But then $(x - z_i)(x - \overline{z_i})$ is a real factor of degree two. ∎

Corollary 3.6.4
An irreducible real polynomial must be of degree 1 or 2.

The proof we presented for the Fundamental Theorem of Algebra will motivate our second proof based on the more general Liouville's theorem. This states that an entire function, that is bounded in the complex plane must be a constant. An entire function is a complex function that has a complex derivative at every point in $\mathbb{C}$. To understand this theorem we must develop some complex analysis, that is, the calculus for complex functions. We will do this in the next two chapters.

Exercises

3.1. Prove Lemma 3.1.1 - that is, if $P(x), Q(x) \in F[x]$, then deg $P(x)Q(x)$ = deg $P(x)$ + deg $Q(x)$ and deg $(P(x) \pm Q(x)) \leq$ Max(deg $P(x)$, deg $Q(x)$).

3.2. Verify that $F[x]$ is a commutative ring by showing that the ring axioms hold.

3.3. Let S be a subring of the field F (such as $\mathbb{Z}$ in $\mathbb{R}$). Let $S[x]$ consist of the polynomials in $F[x]$ with coefficients from S. Show that $S[x]$ is a subring of $F[x]$. Recall that to show a subset is a subring we must only show that it is nonempty and closed under addition, subtraction, and multiplication.

3.4. The following theorem can be proved:
Theorem
Given $(n + 1)$ distinct values $x_0, x_1, \ldots, x_n$ in a field F and $(n + 1)$ other values $y_0, y_1, \ldots, y_n$ in F, then there exists a unique polynomial $f(x) \in F[x]$ of degree $\leq n$ such that $f(x_i) = y_i$ for $i = 0, 1, \ldots, n$.

The proof is from linear algebra. Suppose $f(x) = a_0 + a_1 x + \cdots + a_n x^n$ with the a_i considered as variables. Using $f(x_i) = y_i$ set up the $(n + 1) \times (n + 1)$

system of equations:

$$a_0 + a_1 x_0 + \cdots + a_n x_0^n = y_0$$
$$a_0 + a_1 x_1 + \cdots + a_n x_1^n = y_1$$
$$\cdots$$
$$a_0 + a_1 x_n + \cdots + a_n x_n^n = y_n$$

The matrix of this system (with the a_i as unknowns) is called the **Vandermonde matrix**, and if the x_i are distinct it can be shown to have a nonzero determinant and thus there exists a unique solution for $a_0, a_1, \ldots, a_n$.
(a) Find the unique polynomial $P(x) \in \mathbb{R}[x]$ of deg ≤ 2 that satisfies $P(0) = 1, P(1) = 2, P(2) = 2$.
(b) Find the unique polynomial $P(x) \in \mathbb{Z}_7[x]$ of deg ≤ 2 that satisfies $P(0) = 1, P(1) = 2, P(2) = 2$. $\mathbb{Z}_7$ is the field of integers modulo 7.

3.5. (a) Use the theorem in Exercise 3.4 to prove that if F is an infinite field and $f(x), g(x) \in F[x]$ with $f(s) = g(s)$ for all $s \in F$, then $f(x)$ and $g(x)$ are the same polynomial.
(b) Show this is not necessarily true over a finite field. (Hint: Consider $f(x) = x + 1, g(x) = x^2 + 1$ in $\mathbb{Z}_2[x]$.)

3.6. Use the division algorithm to find the quotient and remainder for the following pairs of polynomials in the indicated polynomial rings.
(a) $f(x) = x^3 + 5x^2 + 6x + 1, g(x) = x - 1$ in $\mathbb{R}[x]$.
(b) $f(x) = x^3 + 5x^2 + 6x + 1, g(x) = x - 1$ in $\mathbb{Z}_5[x]$.
(c) $f(x) = x^3 + 5x^2 + 6x + 1, g(x) = x - 1$ in $\mathbb{Z}_{13}[x]$.

3.7. Use the **Euclidean algorithm** to find the gcd of the following pairs of polynomials in $\mathbb{Q}[x]$.
(a) $f(x) = 2x^3 - 4x^2 + x - 2, g(x) = x^3 - x^2 - x - 2$.
(b) $f(x) = x^4 + x^3 + x^2 + x + 1, g(x) = x^3 - 1$.

3.8. Suppose F is a subfield of K. Then from Exercise 3.3, $F[x]$ is a subring of $K[x]$. Suppose $f(x), g(x), h(x) \in K[x]$ and suppose $f(x) = g(x)h(x)$. Prove: If any two of these are in $F[x]$, then so is the third.

3.9. Show that $f(x) = x^2 + x + 4$ is irreducible in $\mathbb{R}[x]$ but completely factorizes in $\mathbb{C}[x]$ and give its factorization. This same polynomial completely factorizes in $\mathbb{Z}_3[x]$; give its factorization there.

3.10. Formally carry out the derivation of the quadratic formula and show that it holds over any field F of characteristic $\neq 2$ - that is, $2 \neq 0$ in F. (Use the completing the square procedure.) What is the problem with 2?

3.11. Prove part (3) of Lemma 3.4.2: If $P(x), Q(x)$ are complex polynomials, then $\overline{PQ}(x) = \overline{P}(x)\overline{Q}(x)$.

3.12. Show that if $z \in \mathbb{C}$, then $(x - z)(x - \overline{z}) \in \mathbb{R}[x]$.

CHAPTER 4

Complex Analysis and Analytic Functions

4.1 Complex Functions and Analyticity

The proof given in the last section for the Fundamental Theorem of Algebra depended only on the calculus of two-variable real-valued functions. However, the proof suggests a more general result, called Liouville's theorem, which we will develop. From this result the Fundamental Theorem of Algebra will then follow as a simple consequence.

In order to describe this approach we must introduce the basic ideas of **complex analysis**, or **complex variables.** This refers to the area of mathematics that endeavors to extend calculus to complex functions.

A **complex function** $w = f(z)$ is a function $f : \mathbb{C} \to \mathbb{C}$. Here $w, z \in \mathbb{C}$ and are then **complex variables**. Regarding the geometric interpretation of $\mathbb{C}$ as the complex plane, a complex function is then a **mapping**, or **transformation** of the complex plane into itself. If $z = x + iy = (x, y)$, $w = u + iv$, then $u = u(x, y)$ and $v = v(x, y)$ are real-valued two-variable functions. Therefore associated to any complex function are these two real functions.

$$w = f(z) = u(z) + iv(z) \tag{4.1}$$

The function $u(z)$ is called the **real part** of $f(z)$, denoted by $Ref(z)$, and $v(z)$ is the **imaginary part** of $f(z)$, denoted by $Imf(z)$. Analytic (calculus) questions about $f(z)$ will in many cases be referred back to questions about $u(x, y)$ and $v(x, y)$.

4.1. Complex Functions and Analyticity

EXAMPLE 4.1.1
Consider the complex function $f(z) = z^2$. Determine its real and imaginary parts.

Suppose $z = x + iy$. Then $z^2 = (x+iy)^2 = (x^2 - y^2) + i(2xy)$. Hence here $Ref(z) = x^2 - y^2$ while $Imf(z) = 2xy$. □

The essential concepts of calculus – continuity, differentiability and integrability – all depend on limits so our development of complex analysis must start there. However, first, we introduce some necessary ideas from the topology of the complex plane that may be familiar to the reader from advanced calculus.

If $z_0 \in \mathbb{C}$ then an open (circular) ϵ - **neighborhood of z_0**, which we denote by $N_\epsilon(z_0)$, consists of those points within an ϵ distance of z_0.

$$N_\epsilon(z_0) = \{z \in \mathbb{C}; |z - z_0| < \epsilon\}.$$

A **region** is any subset of $\mathbb{C}$. A region $U \subset \mathbb{C}$ is **open** if for each point $z_0 \in U$ there exists some ϵ - neighborhood of z_0 entirely contained in U. A region $C \subset \mathbb{C}$ is **closed** if its complement C' is open. Equivalently, C is closed if for every convergent sequence of points $\{z_n\} \subset C$ with $z_n \to z$, then $z \in C$. A region $U \subset \mathbb{C}$ is **bounded** if U is contained in some disk of radius r centered on the origin; that is, $U \subset \{z; |z| \leq r\}$ for some $r \in \mathbb{R}$. A closed and bounded region in $\mathbb{C}$ is called a **compact region**. Recall from advanced calculus that a real valued function with compact domain D is bounded on D and attains its extreme values (its max and min) on D. Finally, an open region U is **connected** if any two points of U can be connected by a finite sequence of line segments all lying entirely in U, while it is **simply connected** if it is connected and the interior of any simple closed curve lying entirely in U has only points from U. An open connected region in $\mathbb{C}$ is called a **domain**. In figures 4.1 and 4.2 we picture various types of regions.

The next region (Figure 4.2) is not simply connected because closed curves going all around either of the two "holes" would contain points not in the region.

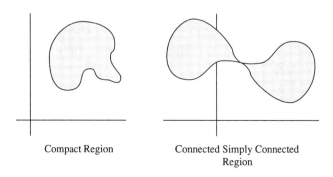

Compact Region   Connected Simply Connected Region

Figure 4.1. Some Regions in $\mathbb{C}$

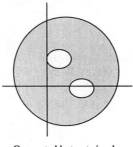

Connected but not simply connected region

Figure 4.2. Some more regions in $\mathbb{C}$

Now we define limits for complex functions in essentially the same manner as for the one-variable real functions of elementary calculus.

Definition 4.1.1
$\lim_{z \to z_0} f(z) = w_0$ if for all $\epsilon > 0$ there exists a $\delta > 0$ such that $|f(z) - w_0| < \epsilon$ whenever $0 < |z - z_0| < \delta$.

Here, of course, the distances $|f(z) - w_0|$ and $|z - z_0|$ are distances in $\mathbb{C}$, and the approach is within a circular neighborhood of z_0. Hence there are infinitely many modes of approach.

All the basic limit theorems from elementary calculus – products, sums, constants, etc. – carry over to this situation. To actually compute limits we pass to the real and imaginary parts.

Lemma 4.1.1
Suppose $f(z) = u(z) + iv(z)$; then
$$\lim_{z \to z_0} f(z) = \lim_{z \to z_0} u(z) + i \lim_{z \to z_0} v(z).$$

EXAMPLE 4.1.2
Let $f(z) = (x^2 + y^2) + i(2xy)$. Find $\lim_{z \to 1+i} f(z)$.
Then
$$\lim_{z \to 1+i} f(z) = \lim_{(x,y) \to (1,1)} (x^2 + y^2) + i \lim_{(x,y) \to (1,1)} (2xy) = 2 + 2i.$$

Using limits we can continue on to continuity and differentiability. □

Definition 4.1.2
$w = f(z)$ is **continuous** at z_0 if $\lim_{z \to z_0} f(z) = f(z_0)$. $f(z)$ is **continuous on a region** U if it is continuous at all points of U.

All the basic results on continuity for one-variable real functions (sums, products, etc.) carry over to complex functions. Further, as with limits,

4.1. Complex Functions and Analyticity

questions of continuity are answered by looking at the real and complex parts.

Lemma 4.1.2
$f(z) = u(z) + iv(z)$ is continuous at $z_0 = (x_0, y_0)$ if and only if $u(x,y), v(x,y)$ are continuous real-valued functions at (x_0, y_0).

Since complex polynomials are built up from only algebraic operations, considered as functions $\mathbb{C} \to \mathbb{C}$, they will be continuous everywhere in $\mathbb{C}$. Since $|z^n| \to \infty$ as $|z| \to \infty$, we have that $|f(z)| \to \infty$ as $|z| \to \infty$ for any nonconstant $f(z) \in \mathbb{C}[z]$. Further, since $|f(z)|$ is then a continuous *real-valued* function, it is bounded on any compact region. We summarize all these facts. From now on we consider a complex polynomial as a polynomial function on $\mathbb{C}$.

Lemma 4.1.3
Let $f(z) \in \mathbb{C}[z]$ then:

(1) $f(z)$ is continuous everywhere in $\mathbb{C}$.
(2) $\lim_{z \to \infty} |f(z)| = \infty$ if $f(z)$ is non-constant.
(3) $|f(z)|$ is bounded on every compact region in $\mathbb{C}$.

We now define the complex derivative in exactly the same formal manner as the real derivative.

Defintion 4.1.3
If $f(z)$ is any complex function, then its **derivative** $f'(z_0)$ at $z_0 \in \mathbb{C}$ is

$$f'(z_0) = \lim_{\Delta z \to 0} \frac{f(z_0 + \Delta z) - f(z_0)}{\Delta z}$$

whenever this limit exists. If $f'(z_0)$ exists, then $f(z)$ is **differentiable** there. $f(z)$ is differentiable on a whole region if it is differentiable at each point of the region.

As with limits and continuity, all the basic differentiation rules from elementary calculus – sums, products, quotients, chain rule etc. – can be shown to be valid for complex functions. In particular the power rule – $f(z) = z^n$ implies $f'(z) = nz^{n-1}$ for n a natural number – follows purely formally from the definition. Therefore, a complex polynomial must have a derivative at each point of $\mathbb{C}$.

Lemma 4.1.4
If $f(z) = a_0 + a_1 z + \cdots + a_n z^n \in \mathbb{C}[z]$ then $f'(z)$ exists at each point $z_0 \in \mathbb{C}$ and $f'(z_0) = a_1 + \cdots + na_n z_0^{n-1}$. Formally, if $f(z) \in \mathbb{C}[z]$ and $\deg f(z) \geq 1$, then $f'(z) \in \mathbb{C}[z]$ and $\deg f'(z) = \deg f(z) - 1$. If $f(z) = a_0$ is constant, then $f'(z) = 0$.

Recall that if $y = f(x)$ is a single-variable real function, then $f'(x_0)$ gives the slope of the tangent at x_0. The complex derivative can also be interpreted geometrically. We will do this in Section 4.3. First we introduce some important general ideas.

Definition 4.1.4
$w = f(z)$ is **analytic** (also called **regular** or **holomorphic**) at z_0 if $f(z)$ is differentiable in a circular neighborhood of z_0. $f(z)$ is analytic in a region U if it is analytic at each point of U. If $f(z)$ is analytic throughout $\mathbb{C}$, then it is called an **entire function**.

From Lemma 4.1.4 it is clear that each complex polynomial is an entire function.

We will develop in the next section conditions on $\text{Re } f(z)$ and $\text{Im } f(z)$ for analyticity. Before this, however, we present an example of a function that has a complex derivative at a point z_0 but that is not analytic at z_0. To understand the example we need the following ideas.

Let $f(z) = u(z) + iv(z)$. Then we define

$$\frac{\partial f}{\partial x} = \frac{\partial u}{\partial x} + i \frac{\partial v}{\partial x} \quad \text{and} \quad \frac{\partial f}{\partial y} = \frac{\partial u}{\partial y} + i \frac{\partial v}{\partial y}.$$

Lemma 4.1.5
Suppose $w = f(z)$ is a real-valued complex function. Then if $f'(z_0)$ exists, $f'(z_0) = \frac{\partial f}{\partial x}(z_0) = \frac{\partial f}{\partial y}(z_0)$.

Proof
From the definition,

$$f'(z_0) = \lim_{\Delta z \to 0} \frac{f(z_0 + \Delta z) - f(z_0)}{\Delta z}.$$

Since $f(z)$ is real-valued, we must have $f(z) = u(z)$ its real part. Then

$$f'(z_0) = \lim_{(\Delta x, \Delta y) \to (0,0)} \frac{u(x_0 + \Delta x, y_0 + \Delta y) - u(x_0, y_0)}{\Delta z}.$$

Since $f'(z_0)$ exists, the limit is independent of the mode of approach. Approaching along a line parallel to the real axis we have $\Delta z = \Delta x$, $\Delta y = 0$. Substituting this in the above we get

$$f'(z_0) = \lim_{\Delta x \to 0} \frac{u(x + \Delta x, y) - u(x, y)}{\Delta x} = \frac{\partial u}{\partial x} = \frac{\partial f}{\partial x}.$$

Similarly approaching along a line parallel to the imaginary axis gives the second part. ∎

EXAMPLE 4.1.3
Let $f(z) = |z|^2$. We show that $f'(0)$ exists but that $f(z)$ is not analytic at $z = 0$.

4.2. The Cauchy-Riemann Equations

Suppose $z_0 = 0$ and $f(z) = |z|^2$ and consider
$$\lim_{\Delta z \to 0} \frac{f(z_0 + \Delta z) - f(z_0)}{\Delta z} = \lim_{\Delta z \to 0} \frac{|\Delta z|^2}{\Delta z} = 0.$$
Therefore, $f'(0)$ exists and $f'(0) = 0$. We show, however, that it cannot be analytic at 0.

Now, if $z = x + iy$, $|z|^2 = x^2 + y^2$. If $f'(z_0)$ exists, then from Lemma 4.1.5, $f'(z_0) = \frac{\partial f}{\partial x}(z_0) = 2x_0 = \frac{\partial f}{\partial y}(z_0) = 2y_0$. This then is possible only if $x_0 = y_0$, and thus the derivative can only possibly exist along the line $y = x$ and hence does not exist in a circular neighborhood of 0. Therefore, it is not analytic at 0. In the next section we will show that this function is differentiable only at 0. □

4.2 The Cauchy-Riemann Equations

Suppose $f(z) = u(z) + iv(z)$ is differentiable at z_0. Then
$$f'(z_0) = \lim_{\Delta z \to 0} \frac{u(z_0 + \Delta z) + iv(z_0 + \Delta z) - (u(z_0) + iv(z_0))}{\Delta z}.$$
Since this limit exists, it is independent of the mode of approach. First allow Δz to approach 0 along a line parallel to the real axis. In this case $\Delta z = \Delta x$ and $\Delta y = 0$. Then if $z_0 = (x_0, y_0)$,
$$f'(z_0) = \lim_{\Delta x \to 0} \frac{u(x_0 + \Delta x, y_0) + iv(x_0 + \Delta x, y_0) - (u(x_0, y_0) + iv(x_0, y_0))}{\Delta x}$$
$$= \lim_{\Delta x \to 0} \frac{u(x_0 + \Delta x, y_0) - u(x_0, y_0)}{\Delta x}$$
$$+ i \lim_{\Delta x \to 0} \frac{v(x_0 + \Delta x, y_0) - v(x_0, y_0)}{\Delta x}$$
$$= \frac{\partial u}{\partial x}(z_0) + i \frac{\partial v}{\partial x}(z_0)$$

Now allow Δz to approach 0 along a line parallel to the imaginary axis. In this case $\Delta z = i\Delta y$ and $\Delta x = 0$. Then
$$f'(z_0) = \lim_{\Delta y \to 0} \frac{u(x_0, y_0 + \Delta y) + iv(x_0, y_0 + \Delta y) - (u(x_0, y_0) + iv(x_0, y_0))}{i\Delta y}$$
$$= \lim_{\Delta y \to 0} \frac{u(x_0, y_0 + \Delta y) - u(x_0, y_0)}{i\Delta y}$$
$$+ i \lim_{\Delta y \to 0} \frac{v(x_0, y_0 + \Delta y) - v(x_0, y_0)}{i\Delta y}$$
$$= \frac{\partial v}{\partial y}(z_0) - i \frac{\partial u}{\partial y}(z_0).$$

Since the derivative exists, these two expressions must be equal. Therefore, at the point z_0 we have

$$\frac{\partial u}{\partial x} = \frac{\partial v}{\partial y} \text{ and } \frac{\partial u}{\partial y} = -\frac{\partial v}{\partial x}. \qquad (4.2.1)$$

These relations are called the **Cauchy-Riemann Equations**, or **Cauchy-Riemann Conditions**, named after A.L. Cauchy who discovered them in the early part of the nineteenth century and B. Riemann who made them fundamental to the theory of complex analysis in the latter part of the same century. Formally, if $u(x, y)$, $v(x, y)$ are real-valued functions:

Definition 4.2.1
$u(x, y)$, $v(x, y)$ satisfy the **Cauchy-Riemann Equations** if

$$\frac{\partial u}{\partial x} = \frac{\partial v}{\partial y} \text{ and } \frac{\partial u}{\partial y} = -\frac{\partial v}{\partial x}.$$

We have thus proved the following theorem.

Theorem 4.2.1
If $f(z) = u(z) + iv(z)$ is differentiable at z_0, then $\frac{\partial u}{\partial x}, \frac{\partial u}{\partial y}, \frac{\partial v}{\partial x}, \frac{\partial v}{\partial y}$ all exist at z_0 and satisfy the Cauchy-Riemann equations. Further,

$$f'(z_0) = \frac{\partial u}{\partial x}(z_0) + i\frac{\partial v}{\partial x}(z_0) = \frac{\partial v}{\partial y}(z_0) - i\frac{\partial u}{\partial y}(z_0).$$

More generally, if $f(z)$ is analytic in some domain U then its real and imaginary parts must satisfy the Cauchy-Riemann equations throughout U. Further, if $f(z) = u(z) + iv(z)$ and $u(z)$, $v(z)$ have continuous partials in U and satisfy the Cauchy-Riemann equations throughout U, then $f(z)$ is analytic in U, which we now prove.

Let $z_0 \in U$. We must then show that $f'(z_0)$ exists. Consider

$$\lim_{\Delta z \to 0} \frac{u(z_0 + \Delta z) + iv(z_0 + \Delta z) - (u(z_0) + iv(z_0))}{\Delta z}$$

$$= \lim_{\Delta z \to 0} \frac{\Delta u + i\Delta v}{\Delta z} = \lim_{\Delta z \to 0} \frac{\Delta u}{\Delta z} + i \lim_{\Delta z \to 0} \frac{\Delta v}{\Delta z}.$$

Since $u(x, y)$, $v(x, y)$ have continuous partials at (x_0, y_0), we have

$$\Delta u = \frac{\partial u}{\partial x}\Delta x + \frac{\partial u}{\partial y}\Delta y + \epsilon_1 \Delta x + \epsilon_2 \Delta y$$

and

$$\Delta v = \frac{\partial v}{\partial x}\Delta x + \frac{\partial v}{\partial y}\Delta y + \epsilon_3 \Delta x + \epsilon_4 \Delta y.$$

4.2. The Cauchy-Riemann Equations

Therefore, using the Cauchy-Riemann equations,

$$\lim_{\Delta z \to 0} \frac{\Delta u + i\Delta v}{\Delta z} =$$

$$\lim_{\Delta z \to 0} \frac{1}{\Delta z} \left(\frac{\partial u}{\partial x}(\Delta x + i\Delta y) + i\frac{\partial v}{\partial x}(\Delta x + i\Delta y) + \delta_1 \Delta x + \delta_2 \Delta y \right).$$

Now, $\Delta z = \Delta x + i\Delta y$, so the above becomes

$$\lim_{\Delta z \to 0} \left(\frac{\partial u}{\partial x} + i\frac{\partial v}{\partial x} + \delta_1 \frac{\Delta x}{\Delta z} + \delta_2 \frac{\Delta y}{\Delta z} \right),$$

where $\delta_1, \delta_2 \to 0$ as $\Delta z \to 0$.

Since $|\Delta x| \leq |\Delta z|$ and $|\Delta y| \leq |\Delta z|$, we have $|\frac{\Delta x}{\Delta z}| \leq 1, |\frac{\Delta y}{\Delta z}| \leq 1$, and hence the final two terms in the limit above tend to zero with Δz. It follows then that at z_0 we must have

$$f'(z_0) = \frac{\partial u}{\partial x}(z_0) + i\frac{\partial v}{\partial x}(z_0),$$

and hence $f(z)$ is differentiable at z_0.

We summarize all these statements in the following theorem and its corollary.

Theorem 4.2.2

(1) *Suppose $f(z) = u(z) + iv(z)$. If $f'(z_0)$ with $z_0 = (x_0, y_0)$ exists, then the partials of $u(x,y), v(x,y)$ must exist at (x_0, y_0) and satisfy the Cauchy-Riemann equations.*

(2) *Suppose $f(z) = u(z) + iv(z)$. If $u(x,y), v(x,y)$ and their partials are continuous at $z_0 = (x_0, y_0)$ and satisfy the Cauchy-Riemann equations at (x_0, y_0), then $f'(z_0)$ exists, that is, $f(z)$ is differentiable at z_0.*

Corollary 4.2.1

Suppose $f(z) = u(z) + iv(z)$ with $u(x,y), v(x,y)$, and their partials continuous on some open domain $U \subset \mathbb{C}$. Then $f(z)$ is analytic on U if and only if u, v satisfy the Cauchy-Riemann equations.

Example 4.2.1

Let $f(z) = e^x \cos y + ie^x \sin y$. Show that $f(z)$ is everywhere analytic and $f'(z) = f(z)$.

Here $u(x,y) = e^x \cos y, v(x,y) = e^x \sin y$. These are everywhere continuous, differentiable, two-variable real-valued functions. Therefore, to show that $f(z)$ is analytic, we must show that they satisfy the Cauchy-Riemann equations.

Now,

$$\frac{\partial u}{\partial x} = e^x \cos y, \quad \frac{\partial u}{\partial y} = -e^x \sin y, \quad \frac{\partial v}{\partial x} = e^x \sin y, \quad \frac{\partial v}{\partial y} = e^x \cos y.$$

Therefore, $\frac{\partial u}{\partial x} = \frac{\partial v}{\partial y}$ and $\frac{\partial v}{\partial x} = -\frac{\partial u}{\partial y}$ for all points in $\mathbb{C}$. Hence $f(z)$ is analytic everywhere in $\mathbb{C}$.

Further, $f'(z) = \frac{\partial u}{\partial x} + i\frac{\partial v}{\partial x} = e^x \cos y + ie^x \sin y = f(z)$. □

The function given in this example is actually the **complex exponential function** $f(z) = e^z$. To see this, suppose $z = x + iy$; then $e^z = e^{x+iy}$ with $x, y \in \mathbb{R}$. Then $e^z = e^x e^{iy} = e^x(\cos y + i \sin y) = e^x \cos y + ie^x \sin y$ from Euler's identity. From this example we see that if $f(z) = e^z$, then $f'(z) = e^z$ also, as we would expect for the exponential function.

EXAMPLE 4.2.2

Using the Cauchy-Riemann equations show that $f(z) = z^2$ is everywhere analytic and $f'(z) = 2z$.

Now, if $z = x + iy$, then $f(z) = z^2 = (x + iy)^2 = x^2 - y^2 + i(2xy)$. Then $u(x, y) = x^2 - y^2$, $v(x, y) = 2xy$. Computing the partial derivatives, we have

$$\frac{\partial u}{\partial x} = 2x, \quad \frac{\partial u}{\partial y} = -2y, \quad \frac{\partial v}{\partial x} = 2y, \quad \frac{\partial v}{\partial y} = 2x.$$

Clearly, these are then continuous and satisfy the Cauchy-Riemann equations throughout $\mathbb{C}$, and hence $f(z)$ is everywhere analytic. Further,

$$f'(z) = \frac{\partial u}{\partial x} + i\frac{\partial v}{\partial x} = 2x + i(2y) = 2(x + iy) = 2z.$$
□

Corollary 4.2.2

(1) The only real-valued analytic functions are constants.

(2) If $f' = 0$ on a region U, then $f(z)$ is a constant.

Proof

(1) Suppose $f(z)$ is real-valued. Then $f(z) = u(z)$ with $v(z) = 0$. If $f(z)$ is analytic, it must satisfy the Cauchy-Riemann equations, so that $\frac{\partial u}{\partial x} = \frac{\partial v}{\partial y} = 0$ and $\frac{\partial u}{\partial y} = -\frac{\partial v}{\partial x} = 0$. Therefore, $\frac{\partial u}{\partial x} = \frac{\partial u}{\partial y} = 0$, and hence $u(x, y)$ is constant, and therefore so is $f(z)$.

(2) If $f' = 0$, then $f' = \frac{\partial u}{\partial x} + i\frac{\partial v}{\partial x} = \frac{\partial v}{\partial y} - i\frac{\partial u}{\partial y} = 0$. This implies that $\frac{\partial u}{\partial x} = \frac{\partial u}{\partial y} = \frac{\partial v}{\partial x} = \frac{\partial v}{\partial y} = 0$, and hence both $u(x, y)$ and $v(x, y)$ are constants. Therefore, $f(z)$ is constant. ■

In example 4.1.2 we showed that although the function $f(z) = |z|^2$ is differentiable at 0 it was not analytic at 0. From the above result we can see that it cannot be analytic anywhere since it is real-valued and not a constant.

4.2. The Cauchy-Riemann Equations

Definition 4.2.2
A real-valued function $u(x, y)$ is a **harmonic function** if it has continuous second partials and satisfies **Laplace's equation**

$$\frac{\partial^2 u}{\partial x^2} + \frac{\partial^2 u}{\partial y^2} = 0.$$

The relevance to our discussion is via the following:

Lemma 4.2.1
If $f(z) = u(z) + iv(z)$ is an analytic function, then $u(x,y)$ and $v(x,y)$ are harmonic functions.

Proof
We leave the continuity of the second partials until later. Suppose $f(z)$ is analytic; then it must satisfy the Cauchy-Riemann equations $\frac{\partial u}{\partial x} = \frac{\partial v}{\partial y}$ and $\frac{\partial u}{\partial y} = -\frac{\partial v}{\partial x}$. Since there are continuous second partials, the order of partial differentiation can be flipped, so that

$$\frac{\partial^2 u}{\partial x^2} = \frac{\partial^2 v}{\partial y \partial x} = \frac{\partial^2 v}{\partial x \partial y} = -\frac{\partial^2 u}{\partial y^2}.$$

Therefore, $\frac{\partial^2 u}{\partial x^2} = -\frac{\partial^2 u}{\partial y^2}$, or $\frac{\partial^2 u}{\partial x^2} + \frac{\partial^2 u}{\partial y^2} = 0$, and so $u(x, y)$ is harmonic. An analogous argument shows that $v(x, y)$ is harmonic. ∎

In the context of the lemma, u, v are called **conjugate harmonic functions.**

EXAMPLE 4.2.3
Show that $u(x, y) = y^3 - 3x^2 y$ is a harmonic function and find a conjugate harmonic function $v(x, y)$ such that $f(z) = u + iv$ is analytic.
Now $\frac{\partial u}{\partial x} = -6xy$, $\frac{\partial u}{\partial y} = 3y^2 - 3x^2$ so $\frac{\partial^2 u}{\partial x^2} = -6y$, $\frac{\partial^2 u}{\partial y^2} = 6y$. Therefore, $\frac{\partial^2 u}{\partial x^2} + \frac{\partial^2 u}{\partial y^2} = 0$, and $u(x, y)$ is harmonic.
Suppose $v(x, y)$ is a conjugate harmonic function. Then from the Cauchy-Riemann equations,

$$\frac{\partial u}{\partial x} = \frac{\partial v}{\partial y}, \quad \frac{\partial u}{\partial y} = -\frac{\partial v}{\partial x}.$$

Consider first $\frac{\partial v}{\partial y} = \frac{\partial u}{\partial x} = -6xy$ and integrate with respect to y to obtain

$$v(x, y) = -3xy^2 + g(x),$$

where $g(x)$ (the constant of integration) is a function of x alone. Then

$$\frac{\partial v}{\partial x} = -3y^2 + g'(x) = -\frac{\partial u}{\partial y} = -3y^2 + 3x^2.$$

This implies that $g'(x) = 3x^2$, and now integrating with respect to x gives $g(x) = x^3 + c$. Any constant will work, so take $c = 0$, and therefore $v(x, y) = x^3 - 3xy^2$ is a conjugate harmonic to $u(x, y)$, and

$$f(z) = (y^3 - 3x^2y) + i(x^3 - 3xy^2)$$

is analytic. □

4.3 Conformal Mappings and Analyticity

Recall from elementary calculus that the derivative $g'(x_0)$ of the single-variable differentiable real function $y = g(x)$ gives the slope of the tangent to the curve $y = g(x)$ at the point $(x_0, g(x_0))$. Thus $g'(x_0)$ gives the following geometric information: first, it gives the direction, or angle, in which the curve is moving at that point, and second, its magnitude gives the instantaneous rate of change of the curve. The complex derivative, and hence analyticity also has a geometric interpretation, which we now discuss.

Definition 4.3.1
A **curve** γ in $\mathbb{C}$ is a continuous function $\gamma : [a, b] \to \mathbb{C}$ given by

$$\gamma(t) = x(t) + iy(t), \qquad (4.3.1)$$

with $x(t)$, $y(t)$ real-valued functions on the real interval $[a, b]$. If $x(t)$, $y(t)$ are differentiable at t_0, then γ is called a **differentiable curve** at t_0 and then $\gamma'(t_0) = x'(t_0) + iy'(t_0)$. A curve is **differentiable** if it is differentiable for all $t \in [a, b]$. A curve is **continuously differentiable** if it is differentiable and the derivative is also continuous on $[a, b]$. The curve $\gamma(t)$ is **regular** at t_0 if $\gamma'(t_0) \neq 0$. The **direction** of $\gamma(t)$ at a regular point t_0 is Arg $\gamma'(t_0)$. In general, a **regular curve** is a curve that is regular at all its points.

EXAMPLE 4.3.1
The curve $\gamma(t) = r \cos t + ir \sin t = re^{it}$, $0 \leq t \leq 2\pi$, represents a circle of radius r centered at the origin.

The derivative is $\gamma'(t) = -r \sin t + ir \cos t$, which is never zero, so $\gamma(t)$ is regular for all t.

At $t = 0$, $\gamma'(0) = ir$. Since this is purely imaginary, the argument is $\pi/2$, as would be expected by looking at the circle in Figure 4.3.

4.3. Conformal Mappings and Analyticity

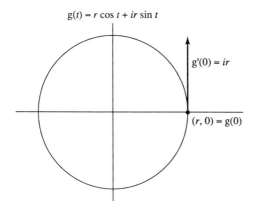

Figure 4.3. A circle in $\mathbb{C}$

More generally, a circle centered at z_0 of radius r is represented by

$$\gamma(t) = z_0 + re^{it}, 0 \leq t \leq 2\pi. \tag{4.3.2}$$

We will need this representation later on.

Suppose γ_1, γ_2 are two curves in $\mathbb{C}$ with $\gamma_1(t_1) = \gamma_2(t_2)$, and both γ_1 and γ_2 are regular at t_1, t_2 respectively. Then the **angle from** γ_1 **to** γ_2 at this common point is Arg $\gamma_2'(t_2)$− Arg $\gamma_1'(t_1)$.

Further, suppose γ is a curve with values in a domain $U \subset \mathbb{C}$, and $F : U \to \mathbb{C}$. Then $F \circ \gamma$ is also a curve. If F has a complex derivative and if γ is differentiable at t_0, then $(F \circ \gamma)'(t_0) = F'(\gamma(t_0))\gamma'(t_0)$. If γ is regular at t_0 and $F'(\gamma(t_0)) \neq 0$, then $F \circ \gamma$ is regular at t_0. □

Definition 4.3.2
Suppose $U \subset \mathbb{C}$ and $F : U \to \mathbb{C}$. Then F is **conformal**, or **isogonal**, at $z_0 \in U$ if for any curve γ, regular at t_0, where $\gamma(t_0) = z_0$, $F \circ \gamma$ is also regular at t_0 and F **preserves angles at** z_0. Preservation of angles means that if $\gamma_1(t), \gamma_2(t)$ are two curves with values in U and $\gamma_1(t_1) = \gamma_2(t_2) = z_0$, then the angle from γ_1 to γ_2 at z_0 is equal to the angle from $F \circ \gamma_1$ to $F \circ \gamma_2$ at $F(z_0)$. If F is conformal throughout U, then it is called a **conformal mapping.**

The relationship between conformality and analyticity is embodied in the next theorem and its corollary.

Theorem 4.3.1
(1) If a continuous complex function $f(z)$ on a domain $U \subset \mathbb{C}$ has a nonzero complex derivative at z_0, then $f(z)$ is conformal at z_0.

(2) Suppose $f(z)$ is continuous on a domain $U \subset \mathbb{C}$ and conformal at $z_0 \in U$, and suppose the partials all exist and are continuous at z_0. Then $f'(z_0)$ exists and is not zero.

Proof

We prove part (1) and leave a sketch of the proof of part (2) to the exercises.

Suppose $f(z)$ is a continuous complex function with a nonzero derivative at z_0. We show that f is conformal at z_0.

Suppose that γ_1, γ_2 are two regular curves with values in U and $\gamma_1(t_1) = \gamma_2(t_2) = z_0$. Since $f'(z_0) \neq 0$, $f \circ \gamma_1, f \circ \gamma_2$ are regular at t_1, t_2. Then the angle from $f \circ \gamma_1$ to $f \circ \gamma_2$ at $f(z_0)$ is

$$\text{Arg}\,((f \circ \gamma_2)'(t_2)) - \text{Arg}\,((f \circ \gamma_1)'(t_1))$$
$$= \text{Arg}\,(f'(\gamma_2(t_2))\gamma_2'(t_2))$$
$$- \text{Arg}\,(f'(\gamma_1(t_1))\gamma_1'(t_1)).$$

Recall that for $z, w \in \mathbb{C}$, $\text{Arg}\,(zw) = \text{Arg}\,z + \text{Arg}\,w$. Therefore, the above becomes

$$\text{Arg}\,(f'(\gamma_2(t_2))) + \text{Arg}\,(\gamma_2'(t_2))) - \text{Arg}\,(f'(\gamma_1(t_1))) - \text{Arg}\,(\gamma_1'(t_1)).$$

However, $\gamma_2(t_2) = \gamma_1(t_1)$, so this in turn becomes

$$\text{Arg}\,(\gamma_2'(t_2)) - \text{Arg}\,(\gamma_1'(t_1)),$$

which is the angle from γ_1 to γ_2 at z_0. Therefore, $f(z)$ preserves angles at z_0 and is thus conformal at z_0. ∎

Corollary 4.3.1
A continuous complex function $f(z)$ for which all the partials exist and are continuous is a conformal mapping on a domain $U \subset \mathbb{C}$ if and only if $f(z)$ is analytic on U and $f'(z) \neq 0$ on U.

Proof

If $f(z)$ is analytic and $f'(z) \neq 0$, then at each point $z_0 \in U$, $f'(z_0) \neq 0$. Then from part (1) of Theorem 4.3.1, $f(z)$ is conformal at z_0 and therefore a conformal mapping on U.

Conversely, suppose $f(z)$ is continuous and conformal at each point of U. Then from part (2) of the theorem $f'(z_0)$ exists and is not zero for each $z_0 \in U$. Therefore, $f(z)$ is analytic on U, and $f'(z) \neq 0$. ∎

Conformality is analogous to the use of the real derivative as a measure of direction. There is a corresponding notion of the instantaneous rate of change for complex functions.

Definition 4.3.3
Suppose $f : U \to \mathbb{C}$, $U \subset \mathbb{C}$, $z_0 \in U$, and $M \geq 0$. Then $f(z)$ is a **magnification** at z_0 by M if

$$\lim_{\Delta z \to 0} \frac{|f(z_0 + \Delta z) - f(z_0)|}{|\Delta z|} = M.$$

Exercises

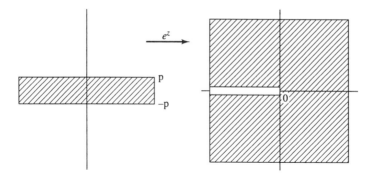

Figure 4.4. e^z as a conformal mapping

Clearly, if $f'(z_0)$ exists, then $f(z)$ is a magnification at z_0 by $|f'(z_0)|$. However, there is a partial converse, which we state as part (2) of the next theorem.

Theorem 4.3.2
(1) If $f(z)$ is differentiable at z_0, then $f(z)$ is a magnification at z_0 by $|f'(z_0)|$.

(2) Let U be a domain in $\mathbb{C}$ and $f(z)$ a continuous complex function on U. Suppose $z_0 \in U, M \geq 0$, and $f(z)$ is a magnification at z_0 by M, and suppose further that at z_0 all the partials exist and are continuous and $f \circ \gamma$ is differentiable at t_0 for any curve γ differentiable at t_0 where $z_0 = \gamma(t_0)$. Then either $f(z)$ is differentiable at z_0, or $\bar{f}(z)$ is differentiable at z_0.

EXAMPLE 4.3.2
Let $f(z) = e^z$. Show that this is a conformal mapping throughout $\mathbb{C}$.

Now, $f(z) = e^z = e^x \cos y + i e^x \sin y$ is never zero in $\mathbb{C}$. From our earlier example we saw that $f'(z) = e^z$ also. Therefore, $f'(z) \neq 0$ in $\mathbb{C}$, and hence it is conformal.

Further, $f(z) = e^z$ maps the strip $-\pi < y < \pi$ one-to-one and conformally onto the split plane that omits 0 and the negative real axis. We indicate this in Figure 4.4. □

Exercises

4.1. Let $f(z) = z^3$.
   (a) Determine the real and imaginary parts of $f(z)$.
   (b) Use the Cauchy-Riemann equations to show that $f(z)$ is everywhere analytic and $f'(z) = 3z^2$.
   (c) Use the formal definition (as in elementary calculus) on z^3 directly to show that $f'(z) = 3z^2$.

4.2. Use induction and the product rule to show that if $f(z) = z^n$, $n \in \mathbb{N}$, then $f'(z) = nz^{n-1}$.

4.3. Show that the function $f(z) = \bar{z}$ is nowhere differentiable.

4.4. Let $f(z)$ be analytic. Show that $g(z) = \overline{f(z)}$ is not analytic unless $f(z)$ is constant.

4.5. Let $f(z) = (2x^2 + y) + i(x^3 - y^3)$.
   (a) What is $f(1 + 2i)$?
   (b) Evaluate $\lim_{z \to 3-i} f(z)$.
   (c) At which points (if any) is $f(z)$ differentiable?
   (d) Is $f(z)$ analytic anywhere? Why?

4.6. Show that the following functions are entire.
   (a) $f(z) = (3x + y) + i(3y - x)$.
   (b) $f(z) = e^{-y} \cos x + ie^{-y} \sin x$.

4.7. Show that the following functions are nowhere analytic.
   (a) $f(z) = xy + iy$.
   (b) $f(z) = e^y \cos x + ie^y \sin x$. Here note the distinction with exercise 4.6 and with the complex exponential function.

4.8. Show that each of the following functions is harmonic and find a harmonic conjugate.
   (a) $u(x, y) = 2x - 2xy$.
   (b) $u(x, y) = 2x - x^3 + 3xy^2$.

4.9. Describe pictorially what the following curves look like.
   (a) $\gamma(t) = (3 + i) + 4e^{it}$, $0 \leq t \leq 2\pi$.
   (b) $\gamma(t) = t + it^2$, $0 \leq t \leq 1$.
   (c) $\gamma(t) = t^2 + i(\ln(t))$, $1 \leq t \leq 2$.

4.10. Let $\gamma_1(t) = t + it^2$, $0 \leq t \leq 1$ and $\gamma_2(t) = \frac{1}{2} + \frac{1}{4}e^{it}$, $0 \leq t \leq 2\pi$. Then $\gamma_1(1/2) = \gamma_2(0) = z_0$. What is the angle between the two curves at this common point?

4.11. (a) Prove formally that if $\lim_{z \to z_0} f(z) = w_1$ and $\lim_{z \to z_0} g(z) = w_2$, then $\lim_{z \to z_0} (f(z) + g(z)) = w_1 + w_2$.
   (b) Use part (a) to show that if $f(z)$ and $g(z)$ are continuous at z_0 then $f(z) + g(z)$ is also continuous at z_0.

4.12. The complex trigonometric functions are defined by
$$\sin z = \frac{e^{iz} - e^{-iz}}{2i}, \quad \cos z = \frac{e^{iz} + e^{-iz}}{2}.$$

(a) Use Euler's identity $e^{it} = \cos t + i \sin t$ applied to a complex variable z to show that these definitions are what you would expect.
(b) Using the derivative of the exponential function find the derivatives, and show that if $f(z) = \sin z$ then $f'(z) = \cos z$, and if $f(z) = \cos z$ then $f'(z) = -\sin z$.

The final exercises will sketch a proof of the second part of Theorem 4.3.1 - that if $f(z)$ is conformal at z_0, then $f(z)$ is differentiable at z_0 and $f'(z_0) \neq 0$.

Exercises

4.13. Define the operator $\frac{\partial}{\partial \bar{z}}$ by

$$\frac{\partial}{\partial \bar{z}} = \frac{1}{2}\left(\frac{\partial}{\partial x} + i\frac{\partial}{\partial y}\right).$$

Let $f(z) = u(z) + iv(z)$ be continuous in a region containing z_0. Suppose at z_0 the partials exist and are continuous. Prove that $f(z)$ being differentiable at z_0 is equivalent to $\frac{\partial f}{\partial \bar{z}} = 0$ at z_0. (Hint: expand out $\frac{\partial f}{\partial \bar{z}}$ and compare to the Cauchy-Riemann equations.)

4.14. For a curve γ regular at t_0 where $\gamma(t_0) = z_0$ and a continuous function $f(z)$ for which the partials exists and are continuous at z_0 and for which $f \circ \gamma$ is also regular at t_0, show that

$$(f \circ \gamma)'(t_0) = \frac{\partial f}{\partial z}(z_0)\gamma'(t_0) + \frac{\partial f}{\partial \bar{z}}(z_0)\overline{\gamma'(t_0)}.$$

Here the operator $\frac{\partial}{\partial z} = \frac{1}{2}(\frac{\partial}{\partial x} - i\frac{\partial}{\partial y})$.

Now we give the proof of the theorem. For each real θ define the curve $\gamma_\theta(t) = z_0 + te^{i\theta}$. Then $z_0 = \gamma_\theta(0)$ for all θ and $\gamma_\theta'(t) = e^{i\theta}$. Further, for all real θ, ϕ, the angle from γ_θ to γ_ϕ at z_0 is $\phi - \theta$.

Since $f(z)$ is conformal at z_0, we have

$$\phi - \theta = \operatorname{Arg}(f \circ \gamma_\phi)'(t_0) - \operatorname{Arg}(f \circ \gamma_\theta)'(t_0).$$

Expanding using Exercise 4.14 and then rearranging, we get that

$$\phi - \theta = \phi - \theta + \operatorname{Arg}\left(\frac{\partial f}{\partial z} + e^{-2i\phi}\frac{\partial f}{\partial \bar{z}}\right) - \operatorname{Arg}\left(\frac{\partial f}{\partial z} + e^{-2i\theta}\frac{\partial f}{\partial \bar{z}}\right).$$

It therefore follows that

$$\operatorname{Arg}\left(\frac{\partial f}{\partial z} + e^{-2i\phi}\frac{\partial f}{\partial \bar{z}}\right) = \operatorname{Arg}\left(\frac{\partial f}{\partial z} + e^{-2i\theta}\frac{\partial f}{\partial \bar{z}}\right).$$

Setting $\theta = 0$, we see then that every circle with center $\frac{\partial f}{\partial z}$ and radius $|\frac{\partial f}{\partial \bar{z}}|$ is contained in the ray $\operatorname{Arg}(z) = \operatorname{Arg}(\frac{\partial f}{\partial z} + \frac{\partial f}{\partial \bar{z}})$. This is possible only if the radius is zero, and hence $|\frac{\partial f}{\partial \bar{z}}| = 0$. Therefore, from Exercise 4.13, $f(z)$ is differentiable.

CHAPTER 5
Complex Integration and Cauchy's Theorem

5.1 Line Integrals and Green's Theorem

In the last chapter we extended differentiation to complex functions. We now move on to an appropriate theory of complex integration. From this theory we will be able to give our second proof of the Fundamental Theorem of Algebra.

Recall that if $z = f(x, y)$ is a real-valued two variable function, there are two appropriate notions of integration. First is the **double integral over a region R**, $\iint_R f(x, y)dA$. This is an integral with respect to area and is an integral of a two-variable function over a two-dimensional object. Alternatively, we use **line integrals** which are integrals of a two-variable function over a a one-dimensional object – a curve. It is the second integral – the line integral – that is most appropriate for complex functions. We use the rest of this section to review the basic ideas of real line integrals.

Definition 5.1.1
Suppose U is a domain in the xy-plane, γ is a curve contained in U, and $f(x, y)$ is a real-valued function defined on U. Choose successive points $P_0 = (x_0, y_0), P_1 = (x_1, y_1), ..., P_n = (x_n, y_n)$ on γ partitioning γ as in Figure 5.1.

Then $f(P_i) = f(x_i, y_i)$ is defined. Let $\Delta x_i = x_i - x_{i-1}$ and form the Riemann sum $\sum_{i=1}^{n} f(x_i, y_i)\Delta x_i$. Then the **line integral of $f(x, y)$ over γ**

5.1. Line Integrals and Green's Theorem

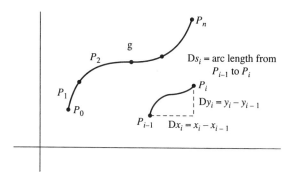

Figure 5.1. Line Integrals

with respect to x, denoted by $\int_\gamma f(x, y)dx$, is defined as

$$\int_\gamma f(x, y)dx = \lim_{\Delta x_i \to 0} \sum_{i=1}^{n} f(x_i, y_i)\Delta x_i \tag{5.1.1}$$

whenever this limit exists independent of the partitioning points.

Analogously, if $\Delta y_i = y_i - y_{i-1}$, the **line integral of $f(x, y)$ over γ with respect to y** is

$$\int_\gamma f(x, y)dy = \lim_{\Delta y_i \to 0} \sum_{i=1}^{n} f(x_i, y_i)\Delta y_i. \tag{5.1.2}$$

Finally, if Δs_i is the arc length along γ from P_{i-1} to P_i, then the **line integral of $f(x, y)$ over γ with respect to arc length** is

$$\int_\gamma f(x, y)ds = \lim_{\Delta s_i \to 0} \sum_{i=1}^{n} f(x_i, y_i)\Delta s_i. \tag{5.1.3}$$

Continuity of $f(x, y)$ over the region U and continuous differentiability of the curve γ are sufficient for the existence of the line integrals.

Lemma 5.1.1
If $f(x,y)$ is continuous on U and $\gamma \subset U$ is continuously differentiable, then the various line integrals exist.

The computation of the various line integrals reverts back to ordinary integration. Suppose $\gamma(t) = x(t) + iy(t) = (x(t), y(t))$ with $t_0 \le t \le t_1$. Then $dx = x'(t)dt$, $dy = y'(t)dt$, $ds = \sqrt{x'(t)^2 + y'(t)^2}dt$. We then obtain by substitution:

$$\int_\gamma f(x, y)dx = \int_{t_0}^{t_1} f(x(t), y(t))x'(t)dt. \tag{5.1.1'}$$

$$\int_\gamma f(x,y)dy = \int_{t_0}^{t_1} f(x(t),y(t))y'(t)dt. \qquad (5.1.2')$$

$$\int_\gamma f(x,y)ds = \int_{t_0}^{t_1} f(x(t),y(t))\sqrt{x'(t)^2 + y'(t)^2}dt. \qquad (5.1.3')$$

Further, if the curve γ is given explicitly as $y = g(x)$ from $x = a$ to $x = b$, then $dx = dx$, $dy = g'(x)dx$, $ds = \sqrt{1 + g'(x)^2}dx$, and the above can be rewritten as:

$$\int_\gamma f(x,y)dx = \int_a^b f(x,g(x))dx. \qquad (5.1.1'')$$

$$\int_\gamma f(x,y)dy = \int_a^b f(x,g(x))g'(x)dx. \qquad (5.1.2'')$$

$$\int_\gamma f(x,y)ds = \int_a^b f(x,g(x))\sqrt{1 + g'(x)^2}dx. \qquad (5.1.3'')$$

For the remainder of this Chapter we will consider only continuously differentiable curves. Hence the word curve is to be interpreted as a continuously differentiable curve.

We illustrate these computations in the next examples.

EXAMPLE 5.1.1
Evaluate the line integral $\int_\gamma 2xy\,dx + (x^2 - y^2)dy$, where γ is the curve given by $x(t) = t^2 - 1$, $y(t) = t^2 + 1$, $0 \leq t \leq 1$.
Here $dx = 2t\,dt$, $dy = 2t\,dt$, so

$$\int_\gamma 2xy\,dx + (x^2 - y^2)dy = \int_0^1 2(t^2 - 1)(t^2 + 1)2t\,dt$$

$$+ \int_0^1 ((t^2 - 1)^2 - (t^2 + 1)^2)2t\,dt = -\frac{10}{3}.$$

□

EXAMPLE 5.1.2
Evaluate the line integral $\int_\gamma (y^3 - 3xy^2)dy$ where γ is the curve given by $y = 2x^2$, $0 \leq x \leq 1$.
Here $dy = 4x\,dx$ so,

$$\int_\gamma (y^3 - 3xy^2)dy = \int_0^1 ((2x^2)^3 - 3x(2x^2)^2)4x\,dx$$

$$= \int_0^1 (32x^7 - 48x^6)dx = -\frac{20}{7}.$$

□

5.1. Line Integrals and Green's Theorem

EXAMPLE 5.1.3
Evaluate the line integral $\int_\gamma y\,ds$, where γ is the curve given by $y = g(x) = \sqrt{x}, 0 \le x \le 6$.
Here $ds = \sqrt{1 + g'(x)^2}\,dx = \frac{1}{2}\sqrt{\frac{1+4x}{x}}\,dx$ so,

$$\int_\gamma y\,ds = \frac{1}{2}\int_0^6 \sqrt{x}\sqrt{\frac{1+4x}{x}}\,dx$$

$$= \frac{1}{2}\int_0^6 (1+4x)^{\frac{1}{2}}\,dx = \frac{31}{3}.$$ □

Line integrals appear in physics most often in the following context. Suppose $\overrightarrow{F(x,y)} = P(x,y)\mathbf{i} + Q(x,y)\mathbf{j}$ is a continuous vector force function acting in the plane. (Here **i** and **j** are the standard unit vectors). Then the work done in pushing a particle along a curve γ subject to this force is

$$W = \int_\gamma P(x,y)\,dx + Q(x,y)\,dy. \qquad (5.1.4)$$

An expression of the form $P(x,y)\,dx + Q(x,y)\,dy$, where $P(x,y), Q(x,y)$ are real-valued functions is called a **first-order differential form**.

Recall that if $f(x,y)$ is a real-valued function with first-order partials, then its **total differential** is

$$df = \frac{\partial f}{\partial x}\,dx + \frac{\partial f}{\partial y}\,dy. \qquad (5.1.5)$$

Thus the total differential of a two-variable function is a first-order differential form. Notice that if γ is a curve with initial point P_0 and terminal point P_1, then by substitution in the definitions,

$$\int_\gamma df = f(P_1) - f(P_0).$$

That is, the line integral of a total differential depends only on the endpoints. We say that $\int_\gamma df$ is **independent of path**, a concept we will return to.

In general, a first-order differential form $P\,dx + Q\,dy$ is **exact** if it is the total differential of some function $f(x,y)$. In this case $P(x,y) = \frac{\partial f}{\partial x}$ and $Q(x,y) = \frac{\partial f}{\partial y}$. Assuming P, Q are continuous we would then have

$$\frac{\partial P}{\partial y} = \frac{\partial^2 f}{\partial x \partial y} = \frac{\partial^2 f}{\partial y \partial x} = \frac{\partial Q}{\partial x}.$$

This condition turns out to be sufficient for exactness as well.

Lemma 5.1.2
If $P(x,y), Q(x,y)$ are continuous, then $P\,dx + Q\,dy$ is exact if and only if $\frac{\partial P}{\partial y} = \frac{\partial Q}{\partial x}$.

The next example illustrates how to determine a function $f(x,y)$ such that $df = Pdx + Qdy$ when $Pdx + Qdy$ is exact. A formal proof of Lemma 5.1.2 is based on this technique.

EXAMPLE 5.1.4
Show that $(x^3 + 3x^2y)dx + (x^3 + y^3)dy$ is exact and determine a function $f(x,y)$ for which it is the total differential.

Here $P(x,y) = (x^3 + 3x^2y)$, $Q(x,y) = (x^3 + y^3)$ and therefore $\frac{\partial P}{\partial y} = 3x^2$, $\frac{\partial Q}{\partial x} = 3x^2$. Hence from Lemma 5.1.2 it is an exact first order differential form.

Suppose

$$df = Pdx + Qdy = (x^3 + 3x^2y)dx + (x^3 + y^3)dy = \frac{\partial f}{\partial x}dx + \frac{\partial f}{\partial y}dy.$$

Then

$$\frac{\partial f}{\partial x} = (x^3 + 3x^2y) \text{ and } \frac{\partial f}{\partial y} = (x^3 + y^3)$$

First integrate $P(x,y) = \frac{\partial f}{\partial x}$ with respect to x to obtain

$$\int (x^3 + 3x^2y)dx = x^4/4 + x^3y + g(y).$$

Therefore, $f(x,y) = x^4/4 + x^3y + g(y)$ where $g(y)$ is a function of y alone.

Differentiating this expression with respect to y it follows that $\frac{\partial f}{\partial y} = x^3 + g'(y) = Q(x,y) = x^3 + y^3$. This implies that $g'(y) = y^3$ and hence $g(y) = y^4/4 + c$. Putting all this together we finally have that

$$f(x,y) = x^4/4 + x^3y + y^4/4 + c.$$

□

We now discuss Green's theorem which may be thought of as a result that expresses the relationship between the two types of integration for two-variable functions (multiple and line). More generally, and outside the scope of these notes, at every dimension (number of variables) there are analagously two types of integrals and a version of Green's theorem.

Definition 5.1.2
A **simple closed curve** is a curve $\gamma(t) : [a,b] \to \mathbb{C}$ such that $\gamma(a) = \gamma(b)$ but $\gamma(t_1) \neq \gamma(t_2)$ for no other pair $t_1, t_2 \in [a,b]$. Basically this is a curve tht is the union of two curves having only their endpoints in common, that is, there is only one point of self-intersection and it occurs at the endpoint of the interval of definition. (See Figure 5.2.)

5.1. Line Integrals and Green's Theorem

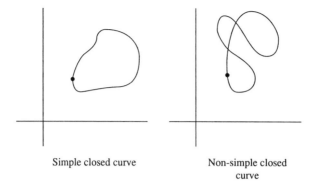

Simple closed curve

Non-simple closed curve

Figure 5.2. Closed Curves

The **Jordan curve theorem** says that a simple closed curve divides the plane into two regions, an interior and an exterior. This seemingly obvious fact is surprisingly difficult to prove.

Theorem 5.1.1
(Green's Theorem) Suppose R is a region in $\mathbb{C}$ whose boundary ∂R is a simple closed curve. Suppose further that $P(x,y), Q(x,y)$ are continuously differentiable functions on a domain U containing R and ∂R. Then

$$\oint_{\partial R} P\,dx + Q\,dy = \int\int_R \left(\frac{\partial Q}{\partial x} - \frac{\partial P}{\partial y}\right) dA,$$

where the line integral is taken around ∂R in the counterclockwise direction.

Green's theorem can be extended to more general regions than those whose boundary is a single simple closed curve. For example if G is a region whose boundary ∂G consists of a finite number of simple closed curves, no two of which intersect, Green's theorem then is still valid – that is

$$\oint_{\partial G} P\,dx + Q\,dy = \int\int_G \left(\frac{\partial Q}{\partial x} - \frac{\partial P}{\partial y}\right) dA,$$

where the line integral over the boundary is defined as the sum over each boundary curve, each directed so that the region G is on the left.

EXAMPLE 5.1.5
Green's theorem applies to the region R in Figure 5.3, where the line integral would be counterclockwise around the outer boundary curve and clockwise around the inner boundary curves. □

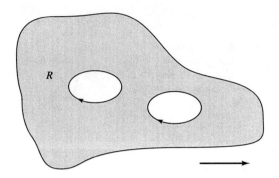

Figure 5.3. Region for Green's Theorem

We prove Green's theorem when the region R is rectangular and then give an example.

Proof

Suppose R is the rectangular region $a \leq x \leq b, c \leq y \leq d$. This is pictured in Figure 5.4.

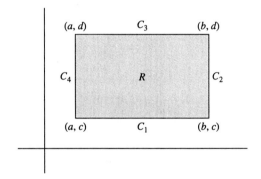

Figure 5.4. Figure 5.4 Rectangular Region

We prove separately that

$$\oint_{\partial R} P(x,y)dx = -\int\int_R \frac{\partial P}{\partial y} dA$$

and

$$\oint_{\partial R} Q(x,y)dy = \int\int_R \frac{\partial Q}{\partial x} dA.$$

Consider first

$$\int\int_R \frac{\partial P}{\partial y} dA = \int_a^b \int_c^d \frac{\partial P}{\partial y} dy dx = \int_a^b (P(x,d) - P(x,c))dx.$$

5.1. Line Integrals and Green's Theorem

Now consider

$$\oint_{\partial R} P(x,y)dx = \int_{C_1} P(x,y)dx + \int_{C_2} P(x,y)dx + \int_{C_3} P(x,y)dx + \int_{C_4} P(x,y)dx$$

where C_1 is the side from (a, c) to (b, c), C_2 is the side from (b, c) to (b, d), C_3 is the side from (b, d) to (a, d), and C_4 is the side from (a, d) to (a, c).

Now, along C_2 and C_4, $dx = 0$, while along C_1, $y = c$ and along C_3, $y = d$. Therefore,

$$\oint_{\partial R} P(x,y)dx = \int_{C_1} P(x,y)dx + \int_{C_3} P(x,y)dx$$

$$= \int_a^b P(x,c)dx + \int_b^a P(x,d)dx = \int_a^b (P(x,c) - P(x,d))dx$$

$$= -\int_a^b (P(x,d) - P(x,c))dx = -\int\int_R \frac{\partial P}{\partial y} dA.$$

An analogous argument works to show that

$$\oint_{\partial R} Q(x,y)dy = \int\int_R \frac{\partial Q}{\partial x} dA.$$

∎

The proof of Green's theorem in general can be accomplished by subdividing the region R into a grid of rectangles.

EXAMPLE 5.1.6
Verify Green's theorem when $P(x,y) = 4y$, $Q(x,y) = 5x$, and R is the region bounded by the unit circle centered on the origin. Hence ∂R is given by $\gamma(t) = e^{it} = \cos t + i \sin t$, $0 \le t \le 2\pi$.
Here $\frac{\partial Q}{\partial x} = 5$, $\frac{\partial P}{\partial y} = 4$, so $\frac{\partial Q}{\partial x} - \frac{\partial P}{\partial y} = 1$. Hence it follows that

$$\int\int_R \left(\frac{\partial Q}{\partial x} - \frac{\partial P}{\partial y} \right) dA = \int\int_R dA = \text{Area}(R) = \pi.$$

Now, ∂R is the unit circle, so $x(t) = \cos t$, $dx = -\sin t\, dt$, $y(t) = \sin t$, $dy = \cos t\, dt$. Hence

$$\oint_{\partial R} P dx + Q dy = \int_0^{2\pi} 4 \sin t(-\sin t) dt + 5 \cos t \cos t\, dt$$

$$= \int_0^{2\pi} (5 \cos^2 t - 4 \sin^2 t) dt = \int_0^{2\pi} (5 - 9 \sin^2 t) dt = \pi.$$
□

Green's theorem and exactness are closely connected to the concept of independence of path. This will play a fundamental role also in complex integration.

Definition 5.1.3
Suppose $P(x,y), Q(x,y)$ are defined on a domain U containing a curve γ. Then $\int_\gamma P dx + Q dy$ is **independent of path** if the value of the integral depends only on the endpoints of γ not on the curve itself.

Theorem 5.1.2
Suppose $P(x,y), Q(x,y)$ are continuously differentiable functions on a domain U. Then the following are equivalent:

(1) $P(x,y)dx + Q(x,y)dy$ is exact.
(2) $\int_\gamma P dx + Q dy$ is independent of path for any curve γ in U.
(3) $\int_\gamma P dx + Q dy = 0$ for any simple closed curve γ in U.

Proof
Suppose $P(x,y)dx + Q(x,y)dy$ is exact. Then there exists a function $f(x,y)$ with $df = P dx + Q dy$. If P_0, P_1 are two points in U and γ is any curve in U with endpoints P_0, P_1, then

$$\int_\gamma P dx + Q dy = \int_\gamma df = f(P_1) - f(P_0).$$

Therefore, $\int_\gamma P dx + Q dy$ is independent of path. Hence (1) implies (2).

Suppose $\int_\gamma P dx + Q dy$ is independent of path for any curve γ in U. Let γ_1 be a simple closed curve. Let P_0, P_1 be two points on γ_1 and let γ_{11} be the curve traversed along γ_1 from P_0 to P_1 and let γ_{12} be the curve traversed along γ_1 from P_0 to P_1 as pictured in Figure 5.5.

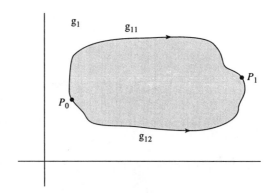

Figure 5.5. Independence of Path

From the independence of path criterion we have

$$\int_{\gamma_{11}} P dx + Q dy = \int_{\gamma_{12}} P dx + Q dy, \text{ or } \int_{\gamma_{11}} P dx + Q dy - \int_{\gamma_{12}} P dx + Q dy = 0.$$

On the other hand, the original curve γ_1 is precisely $\gamma_{11} - \gamma_{12}$, and so

$$\int_{\gamma_1} P dx + Q dy = \int_{\gamma_{11}} P dx + Q dy - \int_{\gamma_{12}} P dx + Q dy = 0.$$

Therefore, (2) implies (3).

Finally, suppose $\int_\gamma P dx + Q dy = 0$ for any simple closed curve γ in U. Then from Green's theorem

$$\iint_{interior\gamma} \left(\frac{\partial Q}{\partial x} - \frac{\partial P}{\partial y} \right) dA = 0.$$

Since this holds for any simple closed curve, it holds for any region in U bounded by a simple closed curve in U. This is possible only if

$$\frac{\partial Q}{\partial x} - \frac{\partial P}{\partial y} = 0 \text{ or } \frac{\partial Q}{\partial x} = \frac{\partial P}{\partial y}.$$

Therefore, $P dx + Q dy$ is exact, completing the proof. ∎

5.2 Complex Integration and Cauchy's Theorem

Using the theory of real line integration we can define the complex integral.

Definition 5.2.1
(a) Suppose $f(t) = u(t) + iv(t)$ is a continuous complex function defined on the interval $t_0 \leq t \leq t_1$. Then we define

$$\int_{t_0}^{t_1} f(t)dt = \int_{t_0}^{t_1} u(t)dt + i \int_{t_0}^{t_1} v(t)dt. \tag{5.2.1}$$

(b) Suppose $f(z) = u(z) + iv(z)$ is a continuous complex function and $\gamma(t) = x(t) + iy(t)$, $t_0 \leq t \leq t_1$, is a curve with $\gamma(t)$ in the domain of $f(z)$. Then $f(\gamma(t))$ is a continuous complex function of the real variable t, and we define the **complex contour integral**, or **complex line integral**, by

$$\int_\gamma f(z)dz = \int_{t_0}^{t_1} f(\gamma(t))\gamma'(t)dt. \tag{5.2.2}$$

This definition of the complex contour integral is independent of the parametrization of the curve in the following sense. Suppose $\gamma_1 : [t_0, t_1] \to \mathbb{C}$ with $t_0 \leq t_1$ and $\gamma_2 : [s_0, s_1] \to \mathbb{C}$ with $s_0 \leq s_1$. Then γ_1 is **equivalent** to γ_2 if there exists a continuously differentiable bijection $\phi : [t_0, t_1] \to [s_0, s_1]$ with $\gamma_2 = \gamma_1 \circ \phi$ and $\phi'(t) > 0$ for all $t \in [t_0, t_1]$. Now if γ_1, γ_2 are two equivalent curves with values in an open region $U \subset \mathbb{C}$ and $f(z)$ is continuous in U, then

$$\int_{\gamma_1} f(z)dz = \int_{\gamma_2} f(z)dz,$$

since $\gamma_2'(t) = \gamma_1'(\phi(t))\phi'(t)$.

If we write $f(z) = u(z) + iv(z)$ and $\gamma(t) = x(t) + iy(t)$, then $\gamma'(t)dt = x'(t)dt + iy'(t)dt = dx + idy = dz$. Substituting, we then have

$$f(z)dz = (u + iv)(dx + idy) = udx - vdy + i(udy + vdx).$$

Therefore, we can rewrite the definition of the complex contour integral as

$$\int_\gamma f(z)dz = \int_\gamma (udx - vdy) + i \int_\gamma (vdx + udy). \quad (5.2.3)$$

EXAMPLE 5.2.1
Evaluate $\int_\gamma z^3 dz$, where γ is the straight line segment from $(0, 0)$ to $(1, 1)$.
Now, $z^3 = (x + iy)^3 = x^3 - 3xy^2 + i(3x^2y - y^3)$, while the straight line segment from $(0, 0)$ to $(1, 1)$ is given by $\gamma(t) = t + it$, $0 \le t \le 1$. Therefore,

$$\int_\gamma f(z)dz = \int_\gamma (udx - vdy) + i \int_\gamma (vdx + udy)$$

$$= \int_\gamma (x^3 - 3xy^2)dx - (3x^2y - y^3)dy + i \int_\gamma (3x^2y - y^3)dx$$
$$+ (x^3 - 3xy^2)dy$$

$$= \int_0^1 (t^3 - 3t^3)dt - (3t^3 - t^3)dt + i \int_0^1 (3t^3 - t^3)dt + (t^3 - 3t^3)dt$$

$$= \int_0^1 -4t^3 dt = -1.$$

$\square$

From the basic results on line integration all the standard integration properties carry over to the complex integral.

Lemma 5.2.1
Let $f(z), g(z)$ be continuous complex functions, $\alpha, \beta \in \mathbb{C}$. Then:

(1) $\int_\gamma \alpha f(z) + \beta g(z) dz = \alpha \int_\gamma f(z)dz + \beta \int_\gamma g(z)dz$.
(2) $|\int_\gamma f(z)dz| \le \int_\gamma |f(z)||dz|$.
(3) $\int_\gamma |dz| = $ arc length of γ.
(4) If $|f(z)| \le M$ on γ and $L = $ arc length of γ, then $|\int_\gamma f(z)dz| \le ML$.

5.2. Complex Integration and Cauchy's Theorem

Suppose $F(z) = U(z) + iV(z)$ is analytic in a region containing a curve γ and suppose further that $f(z) = F'(z)$. Then $f(z) = \frac{\partial U}{\partial x} + i\frac{\partial V}{\partial x}$ and

$$\int_\gamma f(z)dz = \int_\gamma \left(\frac{\partial U}{\partial x}dx - \frac{\partial V}{\partial x}dy\right) + i\int_\gamma \left(\frac{\partial V}{\partial x}dx + \frac{\partial U}{\partial x}dy\right) \qquad (5.2.4)$$

using (5.2.3). From the Cauchy-Riemann equations, $\frac{\partial U}{\partial x} = \frac{\partial V}{\partial y}$ and $\frac{\partial U}{\partial y} = -\frac{\partial V}{\partial x}$, and therefore,

$$\frac{\partial^2 U}{\partial x \partial y} = \frac{\partial^2 U}{\partial y \partial x} = -\frac{\partial^2 V}{\partial x^2} \quad \text{and} \quad \frac{\partial^2 V}{\partial x \partial y} = \frac{\partial^2 V}{\partial y \partial x} = \frac{\partial^2 U}{\partial x^2}.$$

Hence both integrals on the right-hand side of (5.2.4) are independent of path. Therefore, if γ goes from z_0 to z_1, this implies that

$$\int_\gamma f(z)dz = \int_\gamma F'(z)dz = F(z_1) - F(z_0). \qquad (5.2.5)$$

This is the complex version of the fundamental theorem of calculus.

Lemma 5.2.2
Suppose $F(z)$ is analytic in a region U and $f(z) = F'(z)$. Then if γ is any curve in U with endpoints z_0, z_1,

$$\int_\gamma f(z)dz = F(z_1) - F(z_0).$$

Corollary 5.2.1
$\int_\gamma z^n dz = \frac{z^{n+1}}{n+1}\Big|_{z_0}^{z_1} = \frac{z_1^{n+1}}{n+1} - \frac{z_0^{n+1}}{n+1}$ for any natural number n and any curve γ with endpoints z_0, z_1.

EXAMPLE 5.2.2
Evaluate $\int_\gamma z^3 dz$, where γ is the straight line segment from $(0, 0)$ to $(1, 1)$.
Here the endpoints of γ are 0 and $1 + i$, and hence using Corollary 5.2.1 we have

$$\int_\gamma z^3 dz = \frac{z^4}{4}\Big|_0^{1+i} = \frac{(1+i)^4}{4} = -1.$$

Notice that this of course agrees with what we computed directly in Example 5.2.1.

As an immediate consequence of Lemma 5.2.2 we see that

$$\int_\gamma F'(z)dz = 0$$

for any **closed curve** γ. We now generalize this to analytic integrands. Consider

$$\int_\gamma f(z)dz = \int_\gamma (udx - vdy) + i\int_\gamma (vdx + udy)$$

with f(z) analytic in a region U containing γ. From the Cauchy-Riemann equations,

$$\frac{\partial u}{\partial x} = \frac{\partial v}{\partial y}, \frac{\partial u}{\partial y} = -\frac{\partial v}{\partial x}.$$

Therefore, if these partials are continuous in U, it follows that each of the differential forms in the integral above is exact, and hence the line integrals are independent of path. It follows that under these conditions

$$\int_\gamma f(z)dz = 0 \qquad (5.2.6)$$

for any simple closed curve γ contained in U. This result was originally given by Cauchy in the early 1800's. Goursat, later in the century, proved that the analyticity of $f(z)$ alone was sufficient (the proof does not depend on the continuity of the partials and thus is stronger than what can be obtained from Green's theorem). This is then what is termed **Cauchy's theorem** or the **Cauchy-Goursat theorem**. In appendix B we give a formal proof of the extended Goursat result. □

Theorem 5.2.1
(*Cauchy's Theorem*) *Let $f(z)$ be analytic throughout a simply connected domain U and suppose γ is a closed contour entirely contained in U. Then*

$$\int_\gamma f(z)dz = 0.$$

We close this section by proving that Cauchy's theorem implies that every analytic function is itself the derivative of another analytic function.

Theorem 5.2.2
Let $f(z)$ be analytic throughout a simply connected domain U. Then $f(z) = F'(z)$ for some function $F(z)$ analytic in U.

Proof
Let z_0 be a fixed point in U. For any curve γ in U and beginning at z_0 it follows from Cauchy's theorem that the value of the integral $\int_\gamma f(z)dz$ depends only on the endpoint of γ. To see this suppose γ, γ_1 are two curves beginning at z_0 and ending at z_1 as in Figure 5.6.

5.2. Complex Integration and Cauchy's Theorem

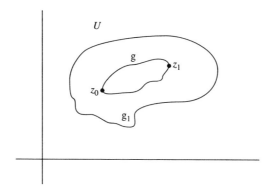

Figure 5.6. Independence of Path

Let γ_2 be the closed path going out along γ and back to z_0 along γ_1. Since $f(z)$ is analytic throughout U we have from Cauchy's theorem

$$\int_{\gamma_2} f(z)dz = 0 = \int_{\gamma} f(z)dz - \int_{\gamma_1} f(z)dz.$$

Hence

$$\int_{\gamma} f(z)dz = \int_{\gamma_1} f(z)dz.$$

Now let $F(z) = \int_{z_0}^{z} f(w)dw$. From the above discussion this is well-defined, and now we show that $F'(z) = f(z)$.

$$F'(z) = \lim_{\Delta z \to 0} \frac{F(z + \Delta z)_F(z)}{\Delta z} = \lim_{\Delta z \to 0} \frac{1}{\Delta z} \left(\int_{z_0}^{z+\Delta z} f(w)dw - \int_{z_0}^{z} f(w)dw \right)$$

$$= \lim_{\Delta z \to 0} \frac{1}{\Delta z} \int_{z}^{z+\Delta z} f(w)dw.$$

Since $f(z)$ is analytic, it is continuous, and then it can be shown (see the exercises) that

$$\int_{z}^{z+\Delta z} f(w)dw = (f(z) + \epsilon)\Delta z,$$

where $\epsilon \to 0$ as $\Delta z \to 0$.

Therefore,

$$\lim_{\Delta z \to 0} \frac{1}{\Delta z} \int_{z}^{z+\Delta z} f(w)dw = f(z),$$

and hence $F'(z) = f(z)$. ∎

5.3 The Cauchy Integral Formula and Cauchy's Estimate

Using Cauchy's theorem we can establish the following fundamental result known as the **Cauchy integral formula**.

Theorem 5.3.1
(Cauchy Integral Formula) Let $f(z)$ be analytic in a simply connected domian U containing a simple closed contour γ. If z_0 is any point interior to γ, then

$$f(z_0) = \frac{1}{2\pi i} \int_\gamma \frac{f(z)}{z - z_0} dz,$$

where the integral is taken in a counterclockwise direction around γ.

Before giving the proof of Theorem 5.3.1, let us note that this theorem implies that the values of a function that is analytic in U within the interior of a curve γ in U are completely determined by the values on the boundary of γ. Thus analyticity forces this extremely strong relationship between values on the boundary of such curves and values in the interior.

Proof
(Theorem 5.3.1) Let C_0 be a circle of radius r_0 centered at z_0. If r_0 is small enough, C_0 will be interior to γ as pictured in Figure 5.7.

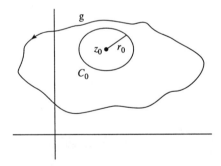

Figure 5.7. Ring-Shaped Region

The function $g(z) = \frac{f(z)}{z-z_0}$ is then analytic in the ring-shaped region bounded by γ and C_0. Hence by Cauchy's theorem

$$\int_\gamma \frac{f(z)}{z - z_0} dz - \int_{C_0} \frac{f(z)}{z - z_0} dz = 0, \qquad (5.3.1)$$

so

$$\int_\gamma \frac{f(z)}{z - z_0} dz = \int_{C_0} \frac{f(z)}{z - z_0} dz.$$

5.3. The Cauchy Integral Formula and Cauchy's Estimate

Then

$$\int_\gamma \frac{f(z)}{z-z_0} dz = f(z_0) \int_{C_0} \frac{dz}{z-z_0} + \int_{C_0} \frac{f(z) - f(z_0)}{z-z_0} dz. \quad (5.3.2)$$

Now, on C_0, $z = z_0 + r_0 e^{it}$ and $dz = ir_0 e^{it} dt$, so

$$\int_{C_0} \frac{dz}{z-z_0} = i \int_0^{2\pi} dt = 2\pi i.$$

Since $f(z)$ is continuous at z_0, if we choose r_0 small enough, we would have $|f(z) - f(z_0)| < \epsilon$ within $|z - z_0| \leq r_0$ (inside C_0). Then

$$\left| \int_{C_0} \frac{f(z) - f(z_0)}{z - z_0} dz \right| \leq \int_{C_0} \frac{|f(z) - f(z_0)|}{|z - z_0|} |dz| < \frac{\epsilon}{r_0} 2\pi r_0 = 2\pi\epsilon.$$

Thus the absolute value of the second integral in (5.3.2) can be made arbitrarily small. Its value must therefore be zero and hence

$$\int_\gamma \frac{f(z)}{z - z_0} dz = 2\pi i f(z_0).$$

∎

EXAMPLE 5.3.1
Evaluate $\int_\gamma \frac{dz}{z^2 - 9}$, where γ is any simple closed contour not containing $z = -3$ in its interior.

Let $f(z) = \frac{1}{z+3}$. If γ does not contain $z = -3$ in its interior, then $f(z)$ is analytic on γ and its interior. Therefore, from the Cauchy integral formula

$$\int_\gamma \frac{dz}{z^2 - 9} = \int_\gamma \frac{dz}{(z+3)(z-3)} = \int_\gamma \frac{f(z)}{z - 3} dz = 2\pi i f(3).$$

Therefore, this value is $2\pi i/6$.

Recall from advanced calculus that if $g(x, t)$ is a differentiable two-variable function and

$$f(t) = \int_a^b g(x, t) dt$$

then

$$f'(t) = \int_a^b \frac{\partial g}{\partial t}(x, t) dt.$$

That is, if a function is defined by integrating a two-variable differentiable function with respect to one variable, then its derivative is the integral of the partial derivative with respect to that variable.

Now let us apply this idea to the Cauchy integral formula to obtain an expression for the derivative of an analytic function. We have

$$f(z_0) = \frac{1}{2\pi i} \int_\gamma \frac{f(z)}{z - z_0} dz.$$

Differentiate inside the integral with respect to z_0 to obtain

$$f'(z_0) = \frac{1}{2\pi i} \int_\gamma \frac{f(z)}{(z - z_0)^2} dz. \qquad (5.3.3)$$

Inductively, we can differentiate this with respect to z_0 to obtain the formula

$$f^{(n)}(z_0) = \frac{n!}{2\pi i} \int_\gamma \frac{f(z)}{(z - z_0)^{n+1}} dz. \qquad (5.3.4)$$

$\square$

Hence we have established the following corollary to the Cauchy integral formula.

Corollary 5.3.1
Let $f(z)$ be analytic in a simply connected domain U containing a simple closed curve γ. If z_0 is interior to γ then an expression for the n – th derivative is

$$f^{(n)}(z_0) = \frac{n!}{2\pi i} \int_\gamma \frac{f(z)}{(z - z_0)^{n+1}} dz.$$

In particular, this shows that if $f(z)$ is analytic in a region U, then it has derivatives of all orders in U.

Corollary 5.3.2
If $f(z)$ is analytic in a simply connected domain U, then it has derivatives of all orders in U.

Recall that an analytic function is the derivative of another analytic function. As a consequence of Corollary 5.3.2 it follows that the derivative of an analytic function is again an analytic function.

Corollary 5.3.3
The derivative of an analytic function is itself analytic.

We note here the contrast with real-valued functions on $\mathbb{R}$. For functions $y = f(x)$ with $x, y \in \mathbb{R}$ it is possible to be differentiable but not twice differentiable. In general, a function $f : [a, b] \to \mathbb{R}$ is of class C^n if it is n-times differentiable. For each n, C^{n+1} is a proper subset of C^n – that is there are n-times differentiable functions that are not $(n + 1)$-times differentiable.

The class C^∞ consists of those functions that are infinitely many times differentiable. Examples include e^x and $\sin x$. Corollary 5.3.2 says that for complex functions, being analytic in a region implies C^∞ in that region.

5.3. The Cauchy Integral Formula and Cauchy's Estimate

Real analytic functions are those functions represented by convergent Taylor series. Clearly, such functions must be C^∞. However, there are C^∞ functions that are not real analytic. This is again not the case for complex functions. Being analytic in a region implies a Taylor series expansion in that region.

Theorem 5.3.2
Let $f(z)$ be analytic at all points within a circle C_0 of radius r_0 centered at z_0. Then within C_0, $f(z)$ is represented by a convergent Taylor series centered at z_0 – that is, for all z within C_0,

$$f(z) = f(z_0) + f'(z_0)(z - z_0) + \ldots + \frac{f^{(n)}(z_0)}{n!}(z - z_0)^n + \ldots.$$

Finally the Cauchy integral formula leads us to **Cauchy's estimate** which will be instrumental in our next proof of the Fundamental Theorem of Algebra.

Theorem 5.3.3
(Cauchy's Estimate) Suppose $f(z)$ is analytic in a simply connected domain U containing the circle C_0 of radius r_0 centered at z_0. If M is the maximum value of $|f(z)|$ on C_0, then

$$|f^{(n)}(z_0)| \leq \frac{n!M}{r_0^n}.$$

In particular, if $|f(z)| < M$ within and on C_0 then

$$|f'(z)| < \frac{M}{r_0}$$

for all z interior to C_0.

Proof
From Corollary 5.3.1

$$|f^{(n)}(z_0)| \leq \left|\frac{n!}{2\pi i} \int_{C_0} \frac{f(z)}{(z-z_0)^{n+1}} dz\right| \leq \frac{Mn!}{2\pi} \left|\int_{C_0} \frac{dz}{(z-z_0)^{n+1}}\right|.$$

On C_0, $z = z_0 + r_0 e^{it}$, $dz = ir_0 e^{it} dt$, so

$$\left|\int_{C_0} \frac{dz}{(z-z_0)^{n+1}}\right| = \left|\int_0^{2\pi} r_0^{-n} e^{-int} dt\right| = 2\pi r_0^{-n}.$$

Hence $|f^{(n)}(z_0)| \leq \frac{Mn!}{r_0^n}$. ∎

5.4 Liouville's Theorem and the Fundamental Theorem of Algebra – Proof Two

Based on Cauchy's estimate we can now prove Liouville's theorem from which we easily obtain our second proof of the Fundamental Theorem of Algebra.

Theorem 5.4.1
(Liouville's Theorem) Suppose $f(z)$ is entire and $|f(z)|$ is bounded for all values of $z \in \mathbb{C}$. Then $f(z)$ is a constant.

(2) More generally, if $|f^{(n)}(z)|$ is bounded throughout $\mathbb{C}$, then $f(z)$ is a polynomial of degree at most $n + 1$.

Proof
Suppose $f(z)$ is entire and $|f(z)| \leq M$ for all $z \in \mathbb{C}$. Then from Cauchy's estimate on a circle of radius r centered at the origin,

$$|f'(z)| < \frac{M}{r}.$$

Since $f(z)$ is entire, we can let $r \to \infty$. Therefore, $|f'(z)| = 0$ and hence $f'(z) = 0$. This of course implies that $f(z)$ must be a constant, completing part (1).

For part (2), if $|f^{(n)}(z)| \leq M$ for all $z \in \mathbb{C}$, then again from Cauchy's estimate

$$|f^{(n+1)}(z)| \leq \frac{M}{r}$$

on any circle of radius r centered at the origin. Again, by letting $r \to \infty$ we would obtain that $f^{(n+1)}(z) = 0$ and hence $f^{(n)}(z)$ is a constant. Then $f(z)$ is a polynomial of degree at most $n + 1$ by antidifferentiation. ∎

Suppose $P(z)$ is a complex polynomial. If deg $P(z) \geq 1$ then $P(z)$ is a nonconstant entire function with the property that $|P(z)| \to \infty$ as $|z| \to \infty$. Combining these facts with Liouville's theorem gives us a proof of the Fundamental Theorem of Algebra.

Theorem 5.4.2
(Fundamental Theorem of Algebra) Let $P(z)$ be a complex polynomial of degree ≥ 1. Then $P(z)$ has at least one complex root.

Proof
Suppose $P(z)$ is a complex polynomial, and let $f(z) = \frac{1}{P(z)}$. If $P(z)$ had no complex root, then $f(z)$ would also be an entire function.

Since $|P(z)| \to \infty$ as $|z| \to \infty$, there exists $M, r > 0$ such that $|P(z)| > M$ if $|z| > r$. This implies that for $|z| > r$, $|f(z)| = \frac{1}{|P(z)|} < \frac{1}{M}$. If $f(z)$ were entire, it would be continuous and thus bounded on the compact set $|z| \le r$. It follows that if $f(z)$ were entire it would be bounded throughout $\mathbb{C}$. From Liouville's theorem it would then follow that $f(z)$ must be a constant. However, then $P(z)$ would also be a constant, which is a contradiction. Therefore, $P(z)$ must be zero for at least one value of $z \in \mathbb{C}$. ∎

What we actually proved above is that if $f(z)$ is a nonconstant entire function with the property that $|f(z)| \to \infty$ as $|z| \to \infty$, then $f(z)$ has at least one zero in $\mathbb{C}$, that is a point $z_0 \in \mathbb{C}$ with $f(z_0) = 0$. However a careful study of entire functions shows that such a function must be a polynomial. Specifically, polynomials can be characterized as entire functions that are infinite at infinity. For more along these lines we refer the reader to the book of Ahlfors [A].

5.5 Some Additional Results

We close this chapter by giving two additional results. While these are not directly relevant to the Fundamental Theorem of Algebra they are part of the same general development.

The first called **Morera's theorem** gives a partial converse to Cauchy's theorem. Specifically if a continuous function has a zero integral over every simple closed curve in a region it must be analytic.

Theorem 5.5.1
(Morera's Theorem) Suppose $f(z)$ is continuous throughout a simply connected domain U and $\int_\gamma f(z)dz = 0$ for every simple closed curve γ interior to U. Then $f(z)$ is analytic throughout U.

Proof
The proof is essentially the same as the proof of Theorem 5.2.2.

Let z_0 be a fixed point in U. For any curve γ_1 beginning at z_0, the value of $\int_{\gamma_1} f(z)dz$ depends only on the endpoint of γ_1 since $\int_\gamma f(z)dz = 0$ for every simple closed curve γ interior to U. Therefore,

$$F(z) = \int_{z_0}^{z} f(w)dw$$

is well-defined. As in the proof of Theorem 5.2.2 we can show that $F'(z) = f(z)$.

Since $F(z)$ is differentiable throughout U, it is analytic and hence $f(z)$ is the derivative of an analytic function. Therefore, from Corollary 5.3.3, $f(z)$ is analytic. ∎

The second result, which we just state, is called the **maximum modulus theorem.** This says that for an analytic function on a compact domain its maximum modulus must occur on the boundary.

Theorem 5.5.2
(*Maximum Modulus Theorem*) *If $f(z)$ is nonconstant and analytic throughout a bounded domain D and continuous on the boundary of D, then $|f(z)|$ assumes its maximum value on the boundary of D, never at an interior point.*

The proof depends on the **maximum-modulus principle**, which states

If $f(z)$ is nonconstant and analytic in a domain U, then very neighborhood in U of $z_0 \in U$ contains points z with $|f(z)| > |f(z_0)|$.

In appendix C we will use the maximum principle to fashion another proof of the Fundamental Theorem of Algebra.

5.6 Concluding Remarks on Complex Analysis

In the past two chapters we have developed many ideas and techniques in complex analysis and used them to produce our second proof of the Fundamental Theorem of Algebra. The key tool in this development was really Cauchy's theorem which began the series of results leading to Liouville's theorem and then our second proof. As mentioned in Section 5.2, Goursat actually proved a much stronger version of Cauchy's theorem. In Appendix B we return to complex analysis and take a more detailed look at Cauchy's theorem and then at the maximum modulus principle. Using this, we then in Appendix C give three additional complex analytic proofs of the Fundamental Theorem of Algebra. In the next two chapters we look at algebraic proofs of this theorem.

Exercises

5.1. Evaluate the following line integrals:
  (a) $\int_\gamma (x^2 + y)dx + (x - y^2)dy$, where γ is the straight line segment from $(0,0)$ to $(2,3)$.

Exercises

(b) $\int_\gamma (x^2 - 2y)dx + (2x + y^2)dy$, where γ is the curve $y^2 = 4x - 1$ from $(1/2, 1)$ to $(3/4, 2)$.

(c) $\int_\gamma \sqrt{x + (3y)^{\frac{5}{3}}}\, ds$, where γ is the curve $y = \frac{1}{3}x^3$ from $x = 0$ to $x = 3$.

(d) $\int_\gamma (x^2 + y^2)dx + (x^2 - y^2)dy$, where γ is the curve $\gamma: x(t) = t^2 + 3, y(t) = t - 1, 1 \le t \le 2$.

(e) $\int_\gamma \frac{ydx + xdy}{\sqrt{x^2 + y^2}}$, where γ is the closed curve $\gamma: x = \cos t, y = \sin t, 0 \le t \le 2\pi$.

5.2. Show that the following integrands are exact first-order differential forms and evaluate the integral by finding an appropriate antiderivative.

(a) $\int_\gamma (x^2 + 2y)dx + (2x + 2y)dy$, where γ is a curve from $(1,1)$ to $(5,3)$.

(b) $\int_\gamma (e^x \cos y)dx - (e^x \sin y)dy$, where γ is a curve from $(0,0)$ to (π, π).

5.3. Verify Green's theorem when

(a) $P(x, y) = -y, Q(x, y) = x$, and R is the unit rectangular region $0 \le x \le 1, 0 \le y \le 1$.

(b) $P(x, y) = xy, Q(x, y) = -2xy$, and R is the rectangular region $1 \le x \le 2, 0 \le y \le 3$.

5.4. Verify (using Green's theorem) that if R is a region bounded by a simple closed curve ∂R then

$$\text{Area}(R) = \int_{\partial R} x\, dy.$$

5.5. Evaluate the following complex integrals:

(a) $\int_\gamma f(z)dz$, where $f(z) = (y - x) - 3x^2 i$ and γ is the straight line segment from $z = 0$ to $z = 1 + i$.

(b) $\int_\gamma f(z)dz$, where $f(z) = \frac{z+2}{z}$ and $\gamma(t) = 2e^{it}, 0 \le t \le \pi$.

5.6. Show directly that

$$\int_{C_0} \frac{dz}{z - z_0} = 2\pi i \text{ and } \int_{C_0} \frac{dz}{(z - z_0)^n} = 0 \text{ if } n = 2, 3, \dots.$$

if C_0 is the circle $\gamma(t) = z_0 + re^{it}, 0 \le t \le 2\pi$. (See how this compares with the Cauchy integral formula.)

5.7. Let γ be the boundary of the square region $-2 \le x \le 2, -2 \le y \le 2$ taken counterclockwise. What are the values of the following integrals and why?

(a) $\int_\gamma \frac{e^z}{z - \pi i/2} dz$.

(b) $\int_\gamma \frac{z^5}{z(z^2+8)} dz$.

(c) $\int_\gamma \frac{z}{2z+1} dz$.

(d) $\int_\gamma \frac{z^4}{(z-(1+i))^2} dz$.

5.8. If $f(z)$ is analytic within and on a curve γ with z_0 in its interior, show that

$$\int_\gamma \frac{f'(z)}{z - z_0} dz = \int_\gamma \frac{f(z)}{(z - z_0)^2} dz.$$

5.9. Show that the real function $f(x) = e^{-\frac{1}{x^2}}$ is C^∞ at $x = 0$ but not real analytic at $x = 0$.

6 Fields and Field Extensions

CHAPTER

6.1 Algebraic Field Extensions

We have now given two proofs of the Fundamental Theorem of Algebra. Both of these involved much more analysis (calculus) than algebra. The first relied on the analytic properties of two-variable real-valued functions from advanced calculus as well as the continuity of real polynomials while the second proof followed from the theory of complex analysis. We now turn to a more algebraic approach to the Fundamental Theorem of Algebra. Eventually we will prove, in the language of this approach, that the complex number field $\mathbb{C}$ is an **algebraically closed field**, a concept equivalent to the fundamental theorem.

If F and F' are fields with F a subfield of F', then F' is an **extension field**, or **field extension**, or simply an **extension**, of F. F' is then a vector space over F (see Chapter 2) and the **degree of the extension** is the dimension of F' as a vector space over F. We denote the degree by $|F':F|$. If the degree is finite, that is, $|F':F| < \infty$, so that F' is a finite-dimensional vector space over F, then F' is called a **finite extension** of F.

From vector space theory we easily obtain that the degrees are multiplicative. Specifically:

Lemma 6.1.1
If $F \subset F' \subset F''$ are fields with F'' a finite extension of F, then $|F':F|$ and $|F'':F'|$ are also finite, and $|F'':F| = |F'':F'||F':F|$.

6.1. Algebraic Field Extensions

Proof
The fact that $|F' : F|$ and $|F'' : F'|$ are also finite follows easily from linear algebra since the dimension of a subspace must be less than the dimension of the whole vector space.

If $|F' : F| = n$ with $\alpha_1, \ldots, \alpha_n$ a basis for F' over F, and $|F'' : F'| = m$ with $\beta_1, \ldots, \beta_m$ a basis for F'' over F' then the mn products $\{\alpha_i \beta_j\}$ form a basis for F'' over F (see the exercises). Then $|F'' : F| = mn = |F'' : F'||F' : F|$. ∎

In the case of the lemma we say that F' is an **intermediate field** (when F and F'' are understood) and F is the **ground field**.

EXAMPLE 6.1.1
(See Example 2.1.3) $\mathbb{C}$ is a finite extension of $\mathbb{R}$, but $\mathbb{R}$ is an infinite extension of $\mathbb{Q}$. □

Our basic approach is to study extension fields whose elements are roots of polynomials over a fixed ground field. To this end we need the following definition.

Definition 6.1.1
Suppose F' is an extension field of F and $\alpha \in F'$. Then α is **algebraic over F** if there exists a polynomial $0 \neq p(x) \in F[x]$ with $p(\alpha) = 0$. (α is a root of a polynomial with coefficients in F.) If every element of F' is algebraic over F, then F' is an **algebraic extension** of F.

If $\alpha \in F'$ is nonalgebraic over F then α is called **transcendental** over F. A nonalgebraic extension is called a **transcendental extension**.

Lemma 6.1.2
Every element of F is algebraic over F.

Proof
If $f \in F$ then $p(x) = x - f \in F[x]$ and $p(f) = 0$. ∎

The tie-in to finite extensions is via the following theorem.

Theorem 6.1.1
If F' is a finite extension of F, then F' is an algebraic extension.

Proof
Suppose $\alpha \in F'$. We must show that there exists a nonzero polynomial $0 \neq p(x) \in F[x]$ with $p(\alpha) = 0$.

Since F' is a finite extension, $|F' : F| = n < \infty$. This implies that there are n elements in a basis for F' over F, and hence any set of $(n+1)$ elements in F' must be linearly dependent over F.

Consider then $1, \alpha, \alpha^2, \ldots, \alpha^n$. These are $(n+1)$ elements in F' and therefore must be linearly dependent. Then there must exist elements $f_0, f_1, \ldots, f_n \in F$ not all zero such that

$$f_0 + f_1\alpha + \cdots + f_n\alpha^n = 0. \tag{6.1.1}$$

Let $p(x) = f_0 + f_1 x + \cdots + f_n x^n$. Then $p(x) \in F[x]$ and from (6.1.1) $p(\alpha) = 0$. ∎

EXAMPLE 6.1.2
$\mathbb{C}$ is algebraic over $\mathbb{R}$, but $\mathbb{R}$ is transcendental over $\mathbb{Q}$.

Since $|\mathbb{C} : \mathbb{R}| = 2$, $\mathbb{C}$ being algebraic over $\mathbb{R}$ follows from theorem 6.1.1. More directly, if $z \in \mathbb{C}$ then $p(x) = (x - z)(x - \bar{z}) \in \mathbb{R}[x]$ and $p(z) = 0$.

$\mathbb{R}$ (and thus $\mathbb{C}$) being transcendental over $\mathbb{Q}$ follows from the existence of transcendental numbers such as e and π (see Chapter 2).

If α is algebraic over F, it satisfies a polynomial over F and hence an irreducible polynomial over F. Since F is a field, if $f \in F$ and $p(x) \in F[x]$, then $f^{-1}p(x) \in F[x]$ also. This implies that if $p(\alpha) = 0$ with a_n the leading coefficient of $p(x)$, then $p_1(x) = a_n^{-1}p(x)$ is a monic polynomial in $F[x]$ that α also satisfies. Thus if α is algebraic over F there is a monic irreducible polynomial that α satisfies. The next result says that this polynomial is unique. □

Lemma 6.1.3
If $\alpha \in F'$ is algebraic over F, then there exists a unique monic irreducible polynomial $p(x) \in F[x]$ such that $p(\alpha) = 0$.

This unique monic irreducible polynomial is denoted by $\mathrm{irr}(\alpha, F)$.

Proof
Suppose $f(\alpha) = 0$ with $0 \neq f(x) \in F[x]$. Then $f(x)$ factors into irreducible polynomials. Since there are no zero divisors in a field, one of these factors, say $p_1(x)$ must also have α as a root. If the leading coefficient of $p_1(x)$ is a_n then $p(x) = a_n^{-1}p_1(x)$ is a monic irreducible polynomial in $F[x]$ that also has α as a root.

Therefore, there exist monic irreducible polynomials that have α as a root. Let $p(x)$ be one such polynomial of minimal degree. It remains to show that $p(x)$ is unique.

Suppose $g(x)$ is another monic irreducible polynomial with $g(\alpha) = 0$. Since $p(x)$ has minimal degree, $\deg p(x) \leq \deg g(x)$. By the division algorithm

$$g(x) = q(x)p(x) + r(x) \tag{6.1.2}$$

where $r(x) \equiv 0$ or $\deg r(x) < \deg p(x)$. Substituting α into (6.1.2) we get

$$g(\alpha) = q(\alpha)p(\alpha) + r(\alpha),$$

6.1. Algebraic Field Extensions

which implies that $r(\alpha) = 0$ since $g(\alpha) = p(\alpha) = 0$. But then if $r(x)$ is not identically 0, α is a root of $r(x)$, which contradicts the minimality of the degree of $p(x)$. Therefore, $r(x) = 0$ and $g(x) = q(x)p(x)$. The polynomial $q(x)$ must be a constant (unit factor) since $g(x)$ is irreducible, but then $q(x) = 1$ since both $g(x), p(x)$ are monic. This says that $g(x) = p(x)$, and hence $p(x)$ is unique. ∎

Suppose $\alpha \in F'$ is algebraic over F and $p(x) = irr(\alpha, F)$. Then there exists a smallest intermediate field E with $F \subset E \subset F'$ such that $\alpha \in E$. By smallest we mean that if E' is another intermediate field with $\alpha \in E'$ then $E \subset E'$. To see that this smallest field exists, notice that there are subfields E' in F' in which $\alpha \in E'$ (namely F' itself). Let E be the intersection all subfields of F' containing α and F. E is a subfield of F' (see the exercises) and E contains both α and F. Further, this intersection is contained in any other subfield containing α and F.

This smallest subfield has a very special form.

Definition 6.1.2
Suppose $\alpha \in F'$ is algebraic over F and $p(x) = irr(\alpha, F) = a_0 + a_1 x + \cdots + a_{n-1} x^{n-1} + x^n$. Let

$$F(\alpha) = \{f_0 + f_1\alpha + \cdots + f_{n-1}\alpha^{n-1}; f_i \in F\}.$$

On $F(\alpha)$ define addition and subtraction componentwise and define multiplication by algebraic manipulation, replacing powers of α higher than α^n by using

$$\alpha^n = -a_0 - a_1\alpha - \cdots - a_{n-1}\alpha^{n-1}.$$

Theorem 6.1.2
$F(\alpha)$ forms a finite algebraic extension of F with $|F(\alpha):F| = \deg irr(\alpha, F)$. $F(\alpha)$ is the smallest subfield of F' that contains the root α. A field extension of the form $F(\alpha)$ for some α is called a **simple extension** of F.

Proof
Recall that $F_{n-1}[x]$ is the set of all polynomials over F of degree $\leq n - 1$ together with the zero polynomial. This set forms a vector space of dimension n over F. As defined in definition 6.1.2, relative to addition and subtraction $F(\alpha)$ is the same as $F_{n-1}[x]$, and thus $F(\alpha)$ is a vector space of dimension $\deg irr(\alpha, F)$ over F and hence an abelian group.

Multiplication is done via multiplication of polynomials, so it is straightforward then that $F(\alpha)$ forms a commutative ring with an identity. We must show that it forms a field. To do this we must show that every nonzero element of $F(\alpha)$ has a multiplicative inverse.

Suppose $0 \neq g(x) \in F[x]$. If $\deg g(x) < n = \deg irr(\alpha, F)$, then $g(\alpha) \neq 0$ since $irr(\alpha, F)$ is the irreducible polynomial of minimal degree that has α as a root.

If $h(x) \in F[x]$ with deg $h(x) \geq n$, then $h(\alpha) = h_1(\alpha)$, where $h_1(x)$ is a polynomial of degree $\leq n - 1$, obtained by replacing powers of α higher than α^n by combinations of lower powers using

$$\alpha^n = -a_0 - a_1\alpha - \cdots - a_{n-1}\alpha^{n-1}.$$

Now suppose $g(\alpha) \in F(\alpha)$, $g(\alpha) \neq 0$. Consider the corresponding polynomial $g(x) \in F[x]$ of degree $\leq n-1$. Since $p(x) = irr(\alpha, F)$ is irreducible, it follows that $g(x)$ and $p(x)$ must be relatively prime, that is, $(g(x), p(x)) = 1$. Therefore, there exist $h(x), k(x) \in F[x]$ such that

$$g(x)h(x) + p(x)k(x) = 1.$$

Substituting α into the above we obtain:

$$g(\alpha)h(\alpha) + p(\alpha)k(\alpha) = 1.$$

However, $p(\alpha) = 0$ and $h(\alpha) = h_1(\alpha) \in F(\alpha)$, so that

$$g(\alpha)h_1(\alpha) = 1.$$

It follows then that in $F(\alpha)$, $h_1(\alpha)$ is the multiplicative inverse of $g(\alpha)$. Since every nonzero element of $F(\alpha)$ has such an inverse $F(\alpha)$ forms a field.

F is contained in $F(\alpha)$ by identifying F with the constant polynomials. Therefore, $F(\alpha)$ is an extension field of F. From the definition of $F(\alpha)$, we have that $\{1, \alpha, \alpha^2, \ldots, \alpha^{n-1}\}$ form a basis, so $F(\alpha)$ has degree n over F. Therefore, $F(\alpha)$ is a finite extension and hence an algebraic extension.

If $F \subset E \subset F'$ and E contains α, then clearly E contains all powers of α since E is a subfield. E then contains $F(\alpha)$, and hence $F(\alpha)$ is the smallest subfield containing both F and α. ∎

EXAMPLE 6.1.3
Consider $p(x) = x^3 - 2$ over $\mathbb{Q}$. This is irreducible over $\mathbb{Q}$ but has the root $\alpha = 2^{1/3} \in \mathbb{R}$. The field $\mathbb{Q}(\alpha) = \mathbb{Q}(2^{1/3})$ is then the smallest subfield of $\mathbb{R}$ that contains $\mathbb{Q}$ and $2^{1/3}$.

Here

$$\mathbb{Q}(\alpha) = \{q_0 + q_1\alpha + q_2\alpha^2; q_i \in \mathbb{Q} \text{ and } \alpha^3 = 2\}.$$

We first give examples of addition and multiplication in $\mathbb{Q}(\alpha)$.
Let $g = 3 + 4\alpha + 5\alpha^2$, $h = 2 - \alpha + \alpha^2$. Then

$$g + h = 5 + 3\alpha + 6\alpha^2$$

and

$$gh = 6 - 3\alpha + 3\alpha^2 + 8\alpha - 4\alpha^2 + 4\alpha^3 + 10\alpha^2 - 5\alpha^3 + 5\alpha^4$$

$$= 6 + 5\alpha + 9\alpha^2 - \alpha^3 + 5\alpha^4.$$

But $\alpha^3 = 2$, so $\alpha^4 = 2\alpha$, and then

$$gh = 6 + 5\alpha + 9\alpha^2 - 2 + 5(2\alpha) = 4 + 15\alpha + 9\alpha^2.$$

6.1. Algebraic Field Extensions

We now show how to find the inverse of h in $\mathbb{Q}(\alpha)$.

Let $h(x) = 2 - x + x^2$, $p(x) = x^3 - 2$. Use the Euclidean algorithm as in Chapter 3 to express 1 as a linear combination of $h(x), p(x)$.

$$x^3 - 2 = (x^2 - x + 2)(x + 1) + (-x - 4),$$

$$x^2 - x + 2 = (-x - 4)(-x + 5) + 22.$$

This implies that

$$22 = (x^2 - x + 2)(1 + (x + 1)(-x + 5)) - ((x^3 - 2)(-x + 5))$$

or

$$1 = \frac{1}{22}[(x^2 - x + 2)(-x^2 + 4x + 6)] - [(x^3 - 2)(-x + 5)].$$

Now substituting α and using that $\alpha^3 = 2$, we have

$$1 = \frac{1}{22}[(\alpha^2 - \alpha + 2)(-\alpha^2 + 4\alpha + 6)],$$

and hence

$$h^{-1} = \frac{1}{22}(-\alpha^2 + 4\alpha + 6).$$

$\square$

Now suppose $\alpha, \beta \in F'$ with both algebraic over F and suppose $irr(\alpha, F) = irr(\beta, F)$. From the construction of $F(\alpha)$ we can see that it would be essentially the same as $F(\beta)$. We now make this idea precise.

Definition 6.1.3
Let F', F'' be extension fields of F. An **F-isomorphism** is an isomorphism $\sigma : F' \to F''$ such that $\sigma(f) = f$ for all $f \in F$. That is, an F-isomorphism is an isomorphism of the extension fields that **fixes each element of the ground field**. If F', F'' are F-isomorphic, we denote this relationship by $F' \cong_F F''$.

Lemma 6.1.4
Suppose $\alpha, \beta \in F'$ are both algebraic over F and suppose $irr(\alpha, F) = irr(\beta, F)$. Then $F(\alpha)$ is F-isomorphic to $F(\beta)$.

Proof
Define the map $\sigma : F(\alpha) \to F(\beta)$ by $\sigma(\alpha) = \beta$ and $\sigma(f) = f$ for all $f \in F$. Allow σ to be a homomorphism - that is preserve addition and multiplication. It follows then that σ maps $f_0 + f_1\alpha + \cdots + f_n\alpha^{n-1} \in F(\alpha)$ to $f_0 + f_1\beta + \cdots + f_n\beta^{n-1} \in F(\beta)$. From this it is straightforward that σ is an F-isomorphism. ∎

If $\alpha, \beta \in F'$ are two algebraic elements over F, we use $F(\alpha, \beta)$ to denote $(F(\alpha))(\beta)$. $F(\alpha, \beta)$ and $F(\beta, \alpha)$ are F-isomorphic so we treat them as the same. We now show that the set of algebraic elements over a ground field is closed under the arithmetic operations and from this obtain that the algebraic elements then form a subfield.

Lemma 6.1.5
If $\alpha, \beta \in F'$ are two algebraic elements over F, then $\alpha \pm \beta, \alpha\beta$, and α/β are also algebraic over F.

Proof
Since α, β are algebraic, the subfield $F(\alpha, \beta)$ will be of finite degree over F and therefore algebraic over F. Now, $\alpha, \beta \in F(\alpha, \beta)$ and since $F(\alpha, \beta)$ is a subfield, it follows that $\alpha \pm \beta, \alpha\beta$, and α/β are also elements of $F(\alpha, \beta)$. Since $F(\alpha, \beta)$ is an algebraic extension of F, each of these elements is algebraic over F. ∎

Theorem 6.1.3
*If F' is an extension field of F, then the set of elements of F' that are algebraic over F forms a subfield. This subfield is called the **algebraic closure of** F **in** F'.*

Proof
Let $A_F(F')$ be the set of algebraic elements over F in F'. $A_F(F') \neq \emptyset$ since it contains F. From the previous lemma it is closed under addition, subtraction, multiplication, and division, and therefore it forms a subfield. ∎

We close this section with a final result, that says that every finite extension is formed by taking successive simple extensions.

Theorem 6.1.4
If F' is a finite extension of F, then there exists a finite set of algebraic elements $\alpha_1, \ldots, \alpha_n$ such that $F' = F(\alpha_1, \ldots, \alpha_n)$.

Proof
Suppose $|F' : F| = k < \infty$. Then F' is algebraic over F. Choose an $\alpha_1 \in F', \alpha_1 \notin F$. Then $F \subset F(\alpha_1) \subset F'$ and $|F' : F(\alpha_1)| < k$. If the degree of this extension is 1, then $F' = F(\alpha_1)$, and we are done. If not, choose an $\alpha_2 \in F', \alpha_2 \notin F(\alpha_1)$. Then as above $F \subset F(\alpha_1) \subset F(\alpha_1, \alpha_2) \subset F'$ with $|F' : F(\alpha_1, \alpha_2)| < |F' : F(\alpha_1)|$. As before, if this degree is one we are done; if not, continue. Since k is finite this process must terminate in a finite number of steps. ∎

6.2 Adjoining Roots to Fields

In the previous section we assumed that we began with an extension field and then considered algebraic elements in that extension. The next result, due to Kronecker, is fundamental because it says that given any irreducible polynomial $f(x) \in F[x]$ we can construct an extension field F' of F in which $f(x)$ has a root.

Theorem 6.2.1
(Kronecker's Theorem) Let F be a field and $f(x) \in F[x]$ an irreducible polynomial over F. Then there exists a finite extension F' of F where $f(x)$ has a root. Further, if α is a root in some extension F'' with $irr(\alpha, F) = f(x)$, then F' is F-isomorphic to $F(\alpha)$.

Proof
To construct the field F' we essentially mimic the construction of $F(\alpha)$ as in the last section.

Suppose $f(x) = a_0 + a_1 x + \cdots + a_n x^n$ with $a_n \neq 0$. Define α to satisfy

$$a_0 + a_1 \alpha + \cdots + a_n \alpha^n = 0.$$

Now define $F' = F(\alpha)$ as in the last section. That is,

$$F(\alpha) = \{f_0 + f_1 \alpha + \cdots + f_{n-1} \alpha^{n-1}; f_i \in F\}.$$

Then on $F(\alpha)$ define addition and subtraction componentwise and define multiplication by algebraic manipulation, replacing powers of α higher than α^n by using

$$\alpha^n = \frac{-a_0 - a_1 \alpha - \cdots - a_{n-1} \alpha^{n-1}}{a_n}.$$

$F' = F(\alpha)$ then forms a field of finite degree over F – the proof being identical to that of the last section. The difference between this construction and the construction in Theorem 6.1.2 is that here α is defined to be the root and we constructed the field around it, whereas in the previous construction α was assumed to satisfy the polynomial and $F(\alpha)$ was an already existing field that contained α. ∎

The field F' constructed above is said to be constructed by **adjoining the root α to F**.

EXAMPLE 6.2.1
Let $f(x) = x^2 + 1 \in \mathbb{R}[x]$. This is irreducible over $\mathbb{R}$. We construct the field in which this has a root.

Let α be an indeterminate with $\alpha^2 + 1 = 0$ or $\alpha^2 = -1$. The extension field $\mathbb{R}(\alpha)$ then has the form

$$\mathbb{R}(\alpha) = \{x + \alpha y; x, y \in \mathbb{R}, \alpha^2 = -1\}.$$

It is clear (see Chapter 2) that this field is $\mathbb{R}$-isomorphic to the complex numbers $\mathbb{C}$, that is, $\mathbb{R}(\alpha) \cong \mathbb{R}(i) \cong \mathbb{C}$. □

The construction of the field $F(\alpha)$ in Theorem 6.2.1 is actually part of a much more general algebraic approach. Although we will not need this general approach further in these notes, we outline it here and then show how it applies to Theorem 6.2.1.

Definition 6.2.1
Let R be a commutative ring (*). A subring $I \subset R$ is an **ideal** if $rI \subset I$ for all $r \in R$.

(*) (Ideals can also be defined for noncommutative rings, but this is not necessary for us here.)

EXAMPLE 6.2.2
In the integers $\mathbb{Z}$, let $n\mathbb{Z} = \{nz; z \in \mathbb{Z}\}$ be the set of all multiples of n. Then $n\mathbb{Z}$ is a subring and if $m \in \mathbb{Z}$, $m(nz_1) = n(mz_1) \in n\mathbb{Z}$, so $n\mathbb{Z}$ is an ideal. □

EXAMPLE 6.2.3
Let R be a commutative ring and let $r \in R$. Let $(r) = \{r_1 r; r_1 \in R\} = $ all multiples of r in R. Then (r) forms an ideal (see the exercises) called the **principal ideal generated by** r.

Note that in this language $n\mathbb{Z}$ is just the principal ideal (n) in $\mathbb{Z}$. □

EXAMPLE 6.2.4
Let R be a commutative ring, $I \subset R$, an ideal and let $r \in R$. Define

$$(r, I) = \{r_1 r + i_1; r_1 \in R, i_1 \in I\}.$$

Then (r, I) forms an ideal (see the exercises) called the **ideal generated by** r **and** I. □

Definition 6.2.2
Let $I \subset R$ be an ideal. A **coset** of I in R is a subset of the form $r + I$ for r a given element in R. We denote the set of all cosets of the ideal $I \subset R$ by R/I.

Lemma 6.2.1
The set R/I of cosets of I in R partition R.

6.2. Adjoining Roots to Fields

Proof
On R define the relation $r_1 \sim r_2$ if $r_1 - r_2 \in I$. This is an equivalence relation, and therefore its equivalence classes partition R. For $r \in R$ the equivalence class $[r]$ is precisely the coset $r + I$. ∎

Given an ideal $I \subset R$ define on R/I:

(1) $(r_1 + I) \pm (r_2 + I) = (r_1 \pm r_2) + I$.
(2) $(r_1 + I)(r_2 + I) = (r_1 r_2) + I$.

The fundamental result, which is a straightforward verification, is then:

Theorem 6.2.2
Given a commutative ring R and an ideal $I \subset R$, then the set of cosets R/I forms a commutative ring under the operations defined above. The coset $0 + I$ is the zero element of R/I, while if R has a multiplicative identity 1 then the coset $1 + I$ is the multiplicative identity for R/I. The ring R/I is called the **quotient ring**, *or* **factor ring** *of R modulo the ideal I.*

EXAMPLE 6.2.5
Consider the ring of integers $\mathbb{Z}$ and the ideal $n\mathbb{Z}$. Let $\mathbb{Z}/n\mathbb{Z}$ be the set of cosets of $n\mathbb{Z}$ in $\mathbb{Z}$.

If $x_1, x_2 \in \mathbb{Z}$, then $x_1 \sim x_2$ if $x_1 - x_2 \in n\mathbb{Z}$, that is if $n|x_1 - x_2$. Therefore, there will only be different cosets for different remainders when dividing by n. It follows that there will be one coset for each $x \in \mathbb{Z}, 0 \le x \le n - 1$. Addition and multiplication of cosets is then done via addition and multiplication modulo n, so that $\mathbb{Z}/n\mathbb{Z} \cong \mathbb{Z}_n$ –
– the integers modulo n. □

Definition 6.2.3
An ideal $I \subset R$ is a **maximal ideal** if $I \ne R$ and $(r, I) = R$ for any $r \in R$, $r \notin I$.

Lemma 6.2.2
Suppose R is a commutative ring with an identity. Then R/I is a field if and only if I is a maximal ideal.

Proof
Suppose I is a maximal ideal. We show that R/I is a field. R/I is a commutative ring with an identity, so we must show that each nonzero element has a multiplicative inverse.

For $r \in R$ let $\bar{r} = r + I$ be its corresponding coset. Now let $\bar{r} \in R/I$, $\bar{r} \ne \bar{0}$. Then $\bar{r} = r + I$ with $r \notin I$ since $\bar{0} = I$ is the zero element of R/I. Since I is maximal, $(r, I) = R$ and therefore $1 \in (r, I)$. It follows that there exist $r_1 \in R$, $i_1 \in I$ such that $r_1 r + i_1 = 1$. In terms of cosets then,

$$r_1 r \in 1 + I \text{ or } \overline{r_1 r} = \bar{1} = 1 \text{ in } R/I.$$

Therefore, $\bar{r}_1$ is the multiplicative inverse of $\bar{r}$ and hence R/I is a field.

Conversely, suppose R/I is a field. We show that I is maximal. Suppose $r \in R$, $r \notin I$. Then $\bar{r} \neq \bar{0}$, so there exists $r_1 \in R$ with $\overline{rr_1} = 1$ in R/I. This implies that $r_1 r \in 1 + I$, or $r_1 r + i_1 = 1$. This implies further that $1 \in (r, I)$. If $s \in R$ then $s.1 = s \in (r, I)$ since (r, I) is an ideal. Hence $R \subset (r, I)$, and since $(r, I) \subset R$ we have $(r, I) = R$. Therefore, I is a maximal ideal. ∎

Now we show how this development relates to the field extension theorem. Suppose F is a field and suppose $f(x) \in F[x]$ is an irreducible polynomial over F. Now, $F[x]$ is a commutative ring with an identity. Consider $(f(x))$, the principal ideal in $F[x]$ generated by $f(x)$.

Suppose $g(x) \notin (f(x))$, so that $g(x)$ is not a multiple of $f(x)$. Since $f(x)$ is irreducible, it follows that $(f(x), g(x)) = 1$. Thus there exist $h(x), k(x) \in F[x]$ with

$$h(x)f(x) + k(x)g(x) = 1.$$

The element on the left is in the ideal $(g(x), (f(x)))$, so the identity, 1, is in this ideal. Therefore, the whole ring $F[x]$ is in this ideal. Since $g(x)$ was arbitrary, this implies that the principal ideal $(f(x))$ is maximal.

Let $F' = F[x]/(f(x))$. From Lemma 6.2.2, F' is a field, and since $F \subset F[x]$, it follows that $F \subset F'$. Let $\bar{x}$ be the coset of x. Then $f(\bar{x}) = \overline{f(x)} = \bar{0}$ in F', so $\bar{x}$ is a root in F'. Here the overbars represent cosets. We have therefore constructed a field F' in which the irreducible polynomial $f(x)$ has a root.

6.3 Splitting Fields

We have just seen that given an irreducible polynomial over a field F we could always find a field extension in which this polynomial has a root. We now push this further to obtain field extensions where a given polynomial has all its roots.

Definition 6.3.1
If $0 \neq f(x) \in F[x]$ and F' is an extension field of F, then $f(x)$ **splits** in F', (F' may be F) if $f(x)$ factors into linear factors in $F'[x]$. Equivalently, this means that all the roots of $f(x)$ are in F'.

F' is a **splitting field** for $f(x)$ over F if F' is the smallest extension field of F in which $f(x)$ splits. (A splitting field for $f(x)$ is the smallest extension field in which $f(x)$ has all its possible roots.)

F' is a **splitting field** over F if it is the splitting field for some finite set of polynomials over F.

6.3. Splitting Fields

Theorem 6.3.1
If $0 \neq f(x) \in F[x]$, then there exists a splitting field for $f(x)$ over F.

Proof
The splitting field is constructed by repeated adjoining of roots. Suppose without loss of generality that $f(x)$ is irreducible of degree n over F. From Theorem 6.2.1 there exists a field F' containing α with $f(\alpha) = 0$. Then $f(x) = (x - \alpha)g(x) \in F'[x]$ with $\deg g(x) = n - 1$. By an inductive argument $g(x)$ has a splitting field and therefore so does $f(x)$. ∎

In the next chapter we will give a further characterization of splitting fields. Here we now return to some ideas introduced in Section 6.1.

Recall that if F' is an extension of F, the set of elements of F' algebraic over F forms a subfield called the algebraic closure of F in F'. More generally, we say that a field F' is **algebraically closed** if every nonconstant polynomial in $F'[x]$ has a root in F'.

Note that in this language the Fundamental Theorem of Algebra says that the complex number field $\mathbb{C}$ is algebraically closed.

The next result gives several clearly equivalent formulations of being algebraically closed.

Theorem 6.3.2
Let F be a field. Then the following are equivalent:

(1) *F is algebraically closed.*
(2) *Every nonconstant polynomial $f(x) \in F[x]$ splits in $F[x]$.*
(3) *F has no proper algebraic extensions, that is there is no algebraic field extension E with $F \subset E$ and $F \neq E$.*

Definition 6.3.2
An extension field F' of F is an **algebraic closure** of F if F' is algebraic over F and F' is algebraically closed.

EXAMPLE 6.3.1
For this example we assume the Fundamental Theorem of Algebra – that is, that $\mathbb{C}$ is algebraically closed. Note that $\mathbb{C}$ is not the algebraic closure of $\mathbb{Q}$ since $\mathbb{C}$ is not algebraic over $\mathbb{Q}$. However, the complex algebraic numbers $A_{\mathbb{C}}$, that is, the set of complex numbers which are algebraic over $\mathbb{Q}$, is the algebraic closure of $\mathbb{Q}$.

To see this, notice that $A_{\mathbb{C}}$ is algebraic over $\mathbb{Q}$ by definition. Now we show that it is algebraically closed. Let $f(x) \in A_{\mathbb{C}}[x]$. If α is a root of $f(x)$, then $\alpha \in \mathbb{C}$, and then α is also algebraic over $\mathbb{Q}$ since each element of $A_{\mathbb{C}}$ is algebraic over $\mathbb{Q}$. Therefore, $\alpha \in A_{\mathbb{C}}$ and $A_{\mathbb{C}}$ is algebraically closed.

More generally, if K is an extension field of F and K is algebraically closed, then the algebraic closure of F in K is the algebraic closure of F. □

Given a polynomial $f(x) \in F[x]$ we have seen that we can construct a splitting field. The next result, whose proof depends on the axiom of choice, indicates that this procedure can be extended to obtain an algebraic closure for any field.

Theorem 6.3.3
Every field F has an algebraic closure, and any two algebraic closures of F are F-isomorphic.

6.4 Permutations and Symmetric Polynomials

To obtain our third proof of the Fundamental Theorem of Algebra we need the concept of a symmetric polynomial. In order to introduce this concept we first review some basic ideas from elementary group theory.

Definition 6.4.1
A **group** G is a set with one binary operation which we will denote by multiplication, such that

(1) The operation is associative, that is, $(g_1 g_2)g_3 = g_1(g_2 g_3)$ for all $g_1, g_2, g_3 \in G$.
(2) There exists an identity for this operation, that is, an element 1 such that $1g = g$ for each $g \in G$.
(3) Each $g \in G$ has an inverse for this operation, that is, for each g there exists a g^{-1} with the property that $gg^{-1} = 1$.

If in addition the operation is commutative, ($g_1 g_2 = g_2 g_1$ for all $g_1, g_2 \in G$), the group G is called an **abelian group**. The **order** of G is the number of elements in G, denoted $|G|$. If $|G| < \infty$, G is a **finite group**. $H \subset G$ is a **subgroup** if H is also a group under the same operation as G. Equivalently, H is a subgroup if $H \neq \emptyset$ and H is closed under the operation and inverses.

Groups most often arise from invertible mappings of a set onto itself. Such mappings are called **permutations**.

Definition 6.4.2
If T is a set, a **permutation** on T is a one-to-one mapping of T onto itself. We denote by S_T the set of all permutations on T.

6.4. Permutations and Symmetric Polynomials

Theorem 6.4.1
*For any set T, S_T forms a group under composition called the **symmetric group** on T. If T, T_1 have the same cardinality (size), then $S_T \cong S_{T_1}$. If T is a finite set with $|T| = n$, then S_T is a finite group and $|S_T| = n!$.*

Proof
If S_T is the set of all permutations on the set T, we must show that composition is an operation on S_T that is associative and has an identity and inverses.

Let $f, g \in S_T$. Then f, g are one-to-one mappings of T onto itself. Consider $f \circ g : T \to T$. If $f \circ g(t_1) = f \circ g(t_2)$, then $f(g(t_1)) = f(g(t_2))$ and $g(t_1) = g(t_2)$, since f is one-to-one. But then $t_1 = t_2$ since g is one-to-one.

If $t \in T$, there exists $t_1 \in T$ with $f(t_1) = t$ since f is onto. Then there exists $t_2 \in T$ with $g(t_2) = t_1$ since g is onto. Putting these together, $f(g(t_2)) = t$, and therefore $f \circ g$ is onto. Therefore, $f \circ g$ is also a permutation and composition gives a valid binary operation on S_T.

The identity function $1(t) = t$ for all $t \in T$ will serve as the identity for S_T, while the inverse function for each permutation will be the inverse. Such unique inverse functions exist since each permutation is a bijection.

Finally, composition of functions is always associative and therefore S_T forms a group.

If T, T_1 have the same cardinality, then there exists a bijection $\sigma : T \to T_1$. Define a map $F : S_T \to S_{T_1}$ in the following manner: if $f \in S_T$, let $F(f)$ be the permutation on T_1 given by $F(f)(t_1) = \sigma(f(\sigma^{-1}(t_1)))$. It is straightforward to verify that F is an isomorphism (see the exercises).

Finally, suppose $|T| = n < \infty$. Then $T = \{t_1, \ldots, t_n\}$. Each $f \in S_T$ can be pictured as

$$f = \begin{pmatrix} t_1 & \cdots & t_n \\ f(t_1) & \cdots & f(t_n) \end{pmatrix}.$$

For t_1 there are n choices for $f(t_1)$. For t_2 there are only $n - 1$ choices since f is one-to-one. This continues down to only one choice for t_n. Using the multiplication principle, the number of choices for f and therefore the size of S_T is

$$n(n-1) \cdots 1 = n!.$$

For a set with n elements we denote S_T by S_n called the **symmetric group on n symbols**. ∎

Example 6.4.1
Write down the six elements of S_3 and give the multiplication table for the group.

Name the three elements 1, 2, 3. The six elements of S_3 are then:

$$1 = \begin{pmatrix} 1 & 2 & 3 \\ 1 & 2 & 3 \end{pmatrix}, a = \begin{pmatrix} 1 & 2 & 3 \\ 2 & 3 & 1 \end{pmatrix}, b = \begin{pmatrix} 1 & 2 & 3 \\ 3 & 1 & 2 \end{pmatrix}$$

$$c = \begin{pmatrix} 1 & 2 & 3 \\ 2 & 1 & 3 \end{pmatrix}, d = \begin{pmatrix} 1 & 2 & 3 \\ 3 & 2 & 1 \end{pmatrix}, e = \begin{pmatrix} 1 & 2 & 3 \\ 1 & 3 & 2 \end{pmatrix}.$$

The multiplication table for S_3 can be written down directly by doing the required composition. For example,

$$ac = \begin{pmatrix} 1 & 2 & 3 \\ 2 & 3 & 1 \end{pmatrix} \begin{pmatrix} 1 & 2 & 3 \\ 2 & 1 & 3 \end{pmatrix} = \begin{pmatrix} 1 & 2 & 3 \\ 3 & 2 & 1 \end{pmatrix} = d.$$

To see this, note that $a : 1 \to 2, 2 \to 3, 3 \to 1$; $c : 1 \to 2, 2 \to 1, 3 \to 3$ and so $ac : 1 \to 3, 2 \to 2, 3 \to 1$.

It is somewhat easier to construct the multiplication table if we make some observations. First, $a^2 = b$ and $a^3 = 1$. Next, $c^2 = 1$, $d = ac$, $e = a^2c$ and finally $ac = ca^2$.

From these relations the following multiplication table can be constructed

	1	a	a^2	c	ac	a^2c
1	1	a	a^2	c	ac	a^2c
a	a	a^2	1	ac	a^2c	c
a^2	a^2	1	a	a^2c	c	ac
c	c	a^2c	ac	1	a^2	a
ac	ac	c	a^2c	a	1	a^2
a^2c	a^2c	ac	c	a^2	a	1

To see this, consider, for example, $(ac)a^2 = a(ca^2) = a(ac) = a^2c$. More generally, we can say that S_3 has a **presentation** given by

$$S_3 = < a, c; a^3 = c^2 = 1, ac = ca^2 > .$$

By this we mean that S_3 is **generated by** a, c, or that S_3 has **generators** a, c and the whole group and its multiplication table can be generated by using the **relations** $a^3 = c^2 = 1$, $ac = ca^2$. □

An important result, the form of which we will see later in our work on extension fields, is the following.

Lemma 6.4.1
Let T be a set and $T_1 \subset T$ a subset. Let H be the subset of S_T that **fixes each element of** T_1 – that is, $f \in H$ if $f(t) = t$ for all $t \in T_1$. Then H is a subgroup.

6.4. Permutations and Symmetric Polynomials

Proof

$H \neq \emptyset$ since $1 \in H$. Now suppose $h_1, h_2 \in H$. Let $t_1 \in T_1$ and consider $h_1 \circ h_2(t_1) = h_1(h_2(t_1))$. Now $h_2(t_1) = t_1$ since $h_2 \in H$, but then $h_1(t_1) = t_1$ since $h_1 \in H$. Therefore, $h_1 \circ h_2 \in H$ and H is closed under composition. If h_1 fixes t_1 then h_1^{-1} also fixes t_1 so H is also closed under inverses and is therefore a subgroup. ∎

We now apply these ideas of permutations to certain polynomials over a field.

Definition 6.4.3

Let $y_1, \ldots, y_n$ be (independent) indeterminates over a field F. A polynomial $f(y_1, \ldots, y_n) \in F[y_1, \ldots, y_n]$ is a **symmetric polynomial** in $y_1, \ldots, y_n$ if $f(y_1, \ldots, y_n)$ is unchanged by any permutation σ of $\{y_1, \ldots, y_n\}$, that is, $f(y_1, \ldots, y_n) = f(\sigma(y_1), \ldots, \sigma(y_n))$.

If $F \subset F'$ are fields and $\alpha_1, \ldots, \alpha_n$ are in F', then we call a polynomial $f(\alpha_1, \ldots, \alpha_n)$ with coefficients in F **symmetric** in $\alpha_1, \ldots, \alpha_n$ if $f(\alpha_1, \ldots, \alpha_n)$ is unchanged by any permutation σ of $\{\alpha_1, \ldots, \alpha_n\}$.

EXAMPLE 6.4.2

Let F be a field and $f_0, f_1 \in F$. Let $h(y_1, y_2) = f_0(y_1 + y_2) + f_1(y_1 y_2)$.

There are two permutations on $\{y_1, y_2\}$, namely $\sigma_1 : y_1 \to y_1, y_2 \to y_2$ and $\sigma_2 : y_1 \to y_2, y_2 \to y_1$.

Applying either one of these two to $\{y_1, y_2\}$ leaves $h(y_1, y_2)$ invariant. Therefore, $h(y_1, y_2)$ is a symmetric polynomial. □

Definition 6.4.4

Let $x, y_1, \ldots, y_n$ be indeterminates over a field F (or elements of an extension field F' over F). Form the polynomial

$$p(x, y_1, \ldots, y_n) = (x - y_1) \ldots (x - y_n).$$

The **ith elementary symmetric polynomial** s_i in $y_1, \ldots, y_n$ for $i = 1, \ldots, n$, is $(-1)^i a_i$, where a_i is the coefficient of x^{n-i} in $p(x, y_1, \ldots, y_n)$.

EXAMPLE 6.4.3

Consider y_1, y_2, y_3. Then

$$p(x, y_1, y_2, y_3) = (x - y_1)(x - y_2)(x - y_3)$$
$$= x^3 - (y_1 + y_2 + y_3)x^2 + (y_1 y_2 + y_1 y_3 + y_2 y_3)x - y_1 y_2 y_3.$$

Therefore, the three elementary symmetric polynomials in y_1, y_2, y_3 over any field are

(1) $s_1 = y_1 + y_2 + y_3$.
(2) $s_2 = y_1 y_2 + y_1 y_3 + y_2 y_3$.
(3) $s_3 = y_1 y_2 y_3$.

In general, the pattern of the last example holds for $y_1, \ldots, y_n$. That is,

$$s_1 = y_1 + y_2 + \cdots + y_n$$
$$s_2 = y_1 y_2 + y_1 y_3 + \cdots + y_{n-1} y_n$$
$$s_3 = y_1 y_2 y_3 + y_1 y_2 y_4 + \cdots + y_{n-2} y_{n-1} y_n$$
$$\vdots$$
$$s_n = y_1 \ldots y_n.$$

$\square$

The importance of the elementary symmetric polynomials is that any symmetric polynomial can be built up from the elementary symmetric polynomials. We make this precise in the next theorem called the **fundamental theorem of symmetric polynomials**. We will use this important result several times, and a complete proof for it will be given in Section 6.7.

Theorem 6.4.2
(Fundamental Theorem of Symmetric Polynomials) If P is a symmetric polynomial in the indeterminates $y_1, \ldots, y_n$ over F, that is, $P \in F[y_1, \ldots, y_n]$ and P is symmetric, then there exists a unique $g \in F[y_1, \ldots, y_n]$ such that $f(y_1, \ldots, y_n) = g(s_1, \ldots, s_n)$. That is, any symmetric polynomial in $y_1, \ldots, y_n$ is a polynomial expression in the elementary symmetric polynomials in $y_1, \ldots, y_n$.

From this theorem we obtain the following two lemmas, which will be crucial in our next proof of the Fundamental Theorem of Algebra.

Lemma 6.4.2
Let $p(x) \in F[x]$ and suppose $p(x)$ has the roots $\alpha_1, \ldots, \alpha_n$ in the splitting field F'. Then the elementary symmetric polynomials in $\alpha_1, \ldots, \alpha_n$ are in F.

Proof
Suppose $p(x) = f_0 + f_1 x + \cdots + f_n x^n \in F[x]$. In $F'[x]$, $p(x)$ splits, with roots $\alpha_1, \ldots, \alpha_n$, and thus in $F'[x]$,

$$p(x) = f_n(x - \alpha_1) \ldots (x - \alpha_n).$$

The coefficients are then $f_n(-1)^i s_i(\alpha_1, \ldots, \alpha_n)$, where the $s_i(\alpha_1, \ldots, \alpha_n)$ are the elementary symmetric polynomials in $\alpha_1, \ldots, \alpha_n$. However, $p(x) \in F[x]$, so each coefficient is in F. It follows then that for each i, $f_n(-1)^i s_i(\alpha_1, \ldots, \alpha_n) \in F$, and hence $s_i(\alpha_1, \ldots, \alpha_n) \in F$ since $f_n \in F$. ∎

Lemma 6.4.3
Let $p(x) \in F[x]$ and suppose $p(x)$ has the roots $\alpha_1, \ldots, \alpha_n$ in the splitting field F'. Suppose further that $g(x) = g(x, \alpha_1, \ldots, \alpha_n) \in F'[x]$. If $g(x)$ is a symmetric polynomial in $\alpha_1, \ldots, \alpha_n$, then $g(x) \in F[x]$.

Proof
If $g(x) = g(x, \alpha_1, \ldots, \alpha_n)$ is symmetric in $\alpha_1, \ldots, \alpha_n$, then from Theorem 6.4.2 it is a symmetric polynomial in the elementary symmetric polynomials in $\alpha_1, \ldots, \alpha_n$. From Lemma 6.4.2 these are in the ground field F, so the coefficients of $g(x)$ are in F. Therefore, $g(x) \in F[x]$. ∎

6.5 The Fundamental Theorem of Algebra – Proof Three

We now present our third proof of the Fundamental Theorem.

Theorem 6.5.1
(*Fundamental Theorem of Algebra*) *Any nonconstant complex polynomial has a complex root. In other words, the complex number field $\mathbb{C}$ is algebraically closed.*

The proof depends on the following four lemmas, three of which we have discussed earlier. The crucial one now is the fourth, which says that any real polynomial must have a complex root.

Lemma 6.5.1
Any odd-degree real polynomial must have a real root.

Proof
Recall that this was Theorem 3.4.1 and was a consequence of the intermediate value theorem.

Suppose $P(x) \in \mathbb{R}[x]$ with $\deg P(x) = n = 2k + 1$ and suppose the leading coefficient $a_n > 0$ (the proof is almost identical if $a_n < 0$). Then

$$P(x) = a_n x^n + \text{(lower terms)}$$

and n is odd. Then,

(1) $\lim_{x \to \infty} P(x) = \lim_{x \to \infty} a_n x^n = \infty$ since $a_n > 0$.
(2) $\lim_{x \to -\infty} P(x) = \lim_{x \to -\infty} a_n x^n = -\infty$ since $a_n > 0$ and n is odd.

From (1), $P(x)$ gets arbitrarily large positively, so there exists an x_1 with $P(x_1) > 0$. Similarly, from (2) there exists an x_2 with $P(x_2) < 0$.

A real polynomial is a continuous real-valued function for all $x \in \mathbb{R}$. Since $P(x_1)P(x_2) < 0$, it follows from the intermediate value theorem that there exists an x_3, between x_1 and x_2, such that $P(x_3) = 0$. ∎

Lemma 6.5.2
Any degree-two complex polynomial must have a complex root.

Proof
Recall that this was Lemma 3.4.1 and was a consequence of the quadratic formula and of the fact that any complex number has a square root.
If $P(x) = ax^2 + bx + c$, $a \neq 0$, then the roots formally are

$$x_1 = \frac{-b + \sqrt{b^2 - 4ac}}{2a}, x_2 = \frac{-b - \sqrt{b^2 - 4ac}}{2a}.$$

From DeMoivre's theorem every complex number has a squareroot, hence x_1, x_2 exist in $\mathbb{C}$. They of course may be the same if $b^2 - 4ac = 0$. ∎

Lemma 6.5.3
If every nonconstant real polynomial has a complex root, then every nonconstant complex polynomial has a complex root.

Proof
This was Theorem 3.4.2 and depended on the concept of the conjugate of a complex polynomial (see Section 3.4).
Let $P(x) \in \mathbb{C}[x]$ and suppose that every nonconstant real polynomial has at least one complex root. Let $H(x) = P(x)\overline{P}(x)$. From Lemma 3.4.3, $H(x) \in \mathbb{R}[x]$. By supposition there exists a $z_0 \in \mathbb{C}$ with $H(z_0) = 0$. Then $P(z_0)\overline{P}(z_0) = 0$, and since $\mathbb{C}$ has no zero divisors, either $P(z_0) = 0$ or $\overline{P}(z_0) = 0$. In the first case z_0 is a root of $P(x)$. In the second case $\overline{P}(z_0) = 0$. Then from Lemma 3.4.2 $\overline{P}(z_0) = \overline{P(\overline{z_0})} = P(\overline{z_0}) = 0$. Therefore, $\overline{z_0}$ is a root of $P(x)$. ∎

Now we come to the crucial lemma.

Lemma 6.5.4
Any nonconstant real polynomial has a complex root.

Proof
Let $f(x) = a_0 + a_1 x + \cdots + a_n x^n \in \mathbb{R}[x]$ with $n \geq 1$, $a_n \neq 0$. The proof is an induction on the degree n of $f(x)$.
Suppose $n = 2^m q$ where q is odd. We do the induction on m. If $m = 0$ then $f(x)$ has odd degree and the theorem is true from Lemma 6.5.1. Assume then that the theorem is true for all degrees $d = 2^k q'$ where $k < m$ and q' is odd. Now assume that the degree of $f(x)$ is $n = 2^m q$.

6.5. The Fundamental Theorem of Algebra – Proof Three

Suppose F' is the splitting field for $f(x)$ over $\mathbb{R}$ in which the roots are $\alpha_1, \ldots, \alpha_n$. This exists from our discussion in section 6.3. We show that at least one of these roots must be in $\mathbb{C}$. (In fact, all are in $\mathbb{C}$ but to prove the lemma we need only show at least one.)

Let $h \in \mathbb{Z}$ and form the polynomial

$$H(x) = \prod_{i<j}(x - (\alpha_i + \alpha_j + h\alpha_i\alpha_j)).$$

This is in $F'[x]$. In forming $H(x)$ we chose pairs of roots $\{\alpha_i, \alpha_j\}$, so the number of such pairs is the number of ways of choosing two elements out of $n = 2^m q$ elements. This is given by

$$\frac{(2^m q)(2^m q - 1)}{2} = 2^{m-1} q(2^m q - 1) = 2^{m-1} q'$$

with q' odd. Therefore, the degree of $H(x)$ is $2^{m-1} q'$.

$H(x)$ is a symmetric polynomial in the roots $\alpha_1, \ldots, \alpha_n$. Since $\alpha_1, \ldots, \alpha_n$ are the roots of a real polynomial, from Lemma 6.4.3 any polynomial in the splitting field symmetric in these roots must be a real polynomial.

Therefore, $H(x) \in \mathbb{R}[x]$ with degree $2^{m-1} q'$. By the inductive hypothesis, then, $H(x)$ must have a complex root. This implies that there exists a pair $\{\alpha_i, \alpha_j\}$ with

$$\alpha_i + \alpha_j + h\alpha_i\alpha_j \in \mathbb{C}.$$

Since h was an arbitrary integer, for any integer h_1 there must exist such a pair $\{\alpha_i, \alpha_j\}$ with

$$\alpha_i + \alpha_j + h_1\alpha_i\alpha_j \in \mathbb{C}.$$

Now let h_1 vary over the integers. Since there are only finitely many such pairs $\{\alpha_i, \alpha_j\}$, it follows that there must be at least two different integers h_1, h_2 such that

$$z_1 = \alpha_i + \alpha_j + h_1\alpha_i\alpha_j \in \mathbb{C} \text{ and } z_2 = \alpha_i + \alpha_j + h_2\alpha_i\alpha_j \in \mathbb{C}.$$

Then $z_1 - z_2 = (h_1 - h_2)\alpha_i\alpha_j \in \mathbb{C}$ and since $h_1, h_2 \in \mathbb{Z} \subset \mathbb{C}$ it follows that $\alpha_i\alpha_j \in \mathbb{C}$. But then $h_1\alpha_i\alpha_j \in \mathbb{C}$, from which it follows that $\alpha_i + \alpha_j \in \mathbb{C}$. Then,

$$p(x) = (x - \alpha_i)(x - \alpha_j) = x^2 - (\alpha_i + \alpha_j)x + \alpha_i\alpha_j \in \mathbb{C}[x].$$

However, $p(x)$ is then a degree-two complex polynomial and so from Lemma 6.5.2 its roots are complex. Therefore, $\alpha_i, \alpha_j \in \mathbb{C}$, and therefore $f(x)$ has a complex root. ∎

It is now easy to give a proof of the Fundamental Theorem of Algebra. From Lemma 6.5.4 every nonconstant real polynomial has a complex root. From Lemma 6.5.3 if every nonconstant real polynomial has a complex root, then every nonconstant complex polynomial has a complex root proving the Fundamental Theorem.

6.6 An Application – the Transcendence of e and π

Recall that an **algebraic number** is an $\alpha \in \mathbb{C}$ that is algebraic over $\mathbb{Q}$, that is, a complex number for which there exists a rational polynomial $0 \neq p(x) \in \mathbb{Q}[x]$ with $p(\alpha) = 0$. We saw in this chapter that the set **A** of algebraic numbers forms a subfield of $\mathbb{C}$. A **transcendental number** is an element of $\mathbb{C} - \mathbf{A}$. In Chapter 2, Exercise 2.7, we proved that the algebraic numbers are countable, and therefore there are uncountably many transcendental numbers. However this was purely an existence proof and to show that any particular complex number is transcendental is extremely difficult. In this section we use the Fundamental Theorem of Algebra and the techniques developed in this chapter, to prove that the important mathematical constants e and π are transcendental.

The existence of transcendental numbers was first proved by Liouville in 1844, who showed that the number $\beta = \sum_{j=1}^{\infty} 10^{-j!}$ is transcendental. Hermite in 1873 proved that e is transcendental, while Lindemann did the same for π in 1882.

First we need some preliminary material. Suppose $\alpha \in \mathbb{C}$ is algebraic, then there exists $0 \neq f(x) \in \mathbb{Q}[x]$ with $f(\alpha) = 0$. Since $\mathbb{Q}$ is a field, we can clearly take $f(x)$ to be monic. Thus there exists an irreducible monic polynomial $p_\alpha(x) \in \mathbb{Q}[x]$ of minimal degree that has α as a root. We call this the **minimal polynomial** of α over $\mathbb{Q}$.

A **primitive polynomial** is an integral polynomial $f(x) = a_n x^n + \cdots + a_0$, $n \geq 1$, $a_n \neq 0$, all $a_i \in \mathbb{Z}$, and $\gcd(a_1, \ldots, a_n) = 1$. If α is algebraic with $f(\alpha) = 0$, $f(x) \in \mathbb{Q}[x]$, then by first multiplying the coefficients of $f(x)$ by a large enough integer to make them all integral and then dividing through by the gcd of the coefficients we can find a primitive polynomial $p(x)$ with $p(\alpha) = 0$. This gives us the following lemma.

Lemma 6.6.1
$\alpha \in \mathbb{C}$ is an algebraic number if and only if there exists a primitive polynomial $p(x) \in \mathbb{Z}[x]$ with $p(\alpha) = 0$.

Suppose $p(x) = x^n + a_{n-1} x^{n-1} + \cdots + a_0$, $n \geq 1$, $a_i \in \mathbb{Q}$, is the minimal polynomial of α over $\mathbb{Q}$. Then by the Fundamental Theorem of Algebra we have a splitting of $p_\alpha(x)$ over $\mathbb{C}$, that is,

$$p_\alpha(x) = (x - \alpha_1) \ldots (x - \alpha_n), \alpha_i \in \mathbb{C} \text{ for } i = 1, \ldots, n.$$

We have $\alpha = \alpha_j$ for one $j \in \{1, \ldots, n\}$. Since $p_\alpha(x)$ is irreducible over $\mathbb{Q}$ we have $p_\alpha(x) = p_{\alpha_i}(x)$ for $i = 1, \ldots, n$ and $\alpha_i \neq \alpha_j$ for $i \neq j$. If some α_i were a multiple, nonsimple root of $p_\alpha(x)$ then α_i would also be a root of the formal derivative $p'_\alpha(x)$ and hence of the greatest common divisor $d(x) \in \mathbb{Q}[x]$ of $p_\alpha(x)$ and $p'_\alpha(x)$. The degree of $d(x)$ is positive but smaller than the degree of $p_\alpha(x)$, since $d(x)$ divides $p'_\alpha(x)$, which contradicts the irreducibility of $p_\alpha(x)$.

The complex numbers $\alpha_1, \ldots, \alpha_n$ are called the **conjugates** of α over $\mathbb{Q}$.

Lemma 6.6.2
If $\alpha \in \mathbb{C}$ is an algebraic number, then its conjugates $\alpha_1, \ldots, \alpha_n$ over $\mathbb{Q}$ are exactly the zeros of an irreducible integral polynomial

$$q_\alpha(x) = b_n x^n + \cdots + b_1 x + b_0 \in \mathbb{Z}[x],$$

with $n \geq 1, b_n > 0$ and $\gcd(b_0, \ldots, b_n) = 1$. Further, $n = $ degree of $p_\alpha(x)$.

The polynomial $q_\alpha(x)$ is called the **entire minimal polynomial** of α over $\mathbb{Q}$.

Note that $q_\alpha(x) = r p_\alpha(x)$ for some $r \in \mathbb{Q}$.

Definition 6.6.1
A complex number $\alpha \in \mathbb{C}$ is an **algebraic integer** if there exists a monic integral polynomial with α as a root. That is, there exists $f(x) \in \mathbb{Z}[x]$ with $f(x) = x^n + b_{n-1} x^{n-1} + \cdots + b_0$, $b_i \in \mathbb{Z}, n \geq 1$, and $f(\alpha) = 0$.

Lemma 6.6.3
If $\alpha \in \mathbb{C}$ is an algebraic integer, then all its conjugates $\alpha_1, \ldots, \alpha_n$ are also algebraic integers.

Proof
Let $f(x) \in \mathbb{Z}[x]$ be a monic polynomial with $f(\alpha) = 0$. Since $p_\alpha(x) = p_{\alpha_i}(x)$, for $i = 1, \ldots, n$ we have $p_{\alpha_i}(x) | f(x)$ for $i = 1, \ldots, n$. Hence $f(\alpha_i) = 0$ for $i = 1, \ldots, n$. ∎

Corollary 6.6.1
If $\alpha \in \mathbb{C}$ is an algebraic integer, then its entire minimal polynomial is monic.

Proof
Let $f(x) \in \mathbb{Z}[x]$ be a monic polynomial with $f(\alpha) = 0$. Then also $f(\alpha_i) = 0$ for all the conjugates. Then there is a primitive integral polynomial $h(x) \in \mathbb{Z}[x]$ and an $r \in \mathbb{Q}$ with $r q_\alpha(x) h(x) = f(x)$. It follows easily that $r = \pm 1$ since $q_\alpha(x)$ and $h(x)$ are primitive and $f(x)$ is monic. Hence $q_\alpha(x)$ is also monic. ∎

If $\alpha \in \mathbb{C}$ is an algebraic integer and $\alpha_1, \ldots, \alpha_n$ are its conjugates, then

$$(x - \alpha_1) \cdots (x - \alpha_n) = x^n - s_1 x^{n-1} + \cdots + (-1)^n s_n \in \mathbb{Z}[x];$$

where $s_i = s_i(\alpha_1, \ldots, \alpha_n)$ is the ith elementary symmetric polynomial in $\alpha_1, \ldots, \alpha_n$ (see the previous section). It follows that the elementary symmetric polynomials must be integers. From the main theorem on symmetric polynomials it follows that any symmetric polynomial in the conjugates of an algebraic integer must be an integer. From this we get the following, which extends to algebraic integers the closure properties of general algebraic numbers.

Lemma 6.6.4
Suppose $\alpha, \beta \in \mathbb{C}$ are algebraic integers. Then so are $\alpha \pm \beta$ and $\alpha\beta$.

Proof
Let $\alpha_1 = \alpha, \ldots, \alpha_n$ be the conjugates of α and $\beta_1 = \beta, \ldots, \beta_m$ the conjugates of β. Let

$$f(x) = \prod_{i=1}^{n}\prod_{j=1}^{m}(x - (\alpha_i + \beta_j)) = x^{n+m} + d_{n+m-1}x^{n+m-1} + \cdots + d_0.$$

The coefficients d_k are symmetric functions in α_i, β_j, and therefore from the remarks above we have $d_k \in \mathbb{Z}$. (Notice the similarity to the argument that we used in Lemma 6.5.4). Therefore, $\alpha + \beta$ is an algebraic integer. We treat $\alpha - \beta$ and $\alpha\beta$ analogously. ∎

Corollary 6.6.2
The set of algebraic integers forms a subring of $\mathbb{C}$. Further, the field of algebraic numbers is the quotient field of the ring of algebraic integers.

Theorem 6.6.1
e is a transcendental number, that is transcendental, over $\mathbb{Q}$.

Proof
Let $f(x) \in \mathbb{R}[x]$ with the degree of $f(x) = m \geq 1$. Let $z_1 \in \mathbb{C}, z_1 \neq 0$, and $\gamma : [0, 1] \to \mathbb{C}, \gamma(t) = tz_1$. Let

$$I(z_1) = \int_\gamma e^{z_1 - z}f(z)dz = \left(\int_0^{z_1}\right)_\gamma e^{z_1 - z}f(z)dz.$$

By $(\int_0^{z_1})_\gamma$ we mean the integral from 0 to z_1 along γ. Recall that

$$\left(\int_0^{z_1}\right)_\gamma e^{z_1 - z}f(z)dz = -f(z_1) + e^{z_1}f(0) + \left(\int_0^{z_1}\right)_\gamma e^{z_1 - z}f'(z)dz.$$

It follows then by repeated partial integration that
(1) $I(z_1) = e^{z_1}\sum_{j=1}^{m} f^{(j)}(0) - \sum_{j=0}^{m} f^{(j)}(z_1)$.
Let $|f|(x)$ be the polynomial that we get if we replace the coefficients of $f(x)$ by their absolute values. Since $|e^{z_1-z}| \leq e^{|z_1-z|} \leq e^{|z_1|}$, we get
(2) $|I(z_1)| \leq |z_1|e^{|z_1|}|f|(|z_1|)$.
Now assume that e is an algebraic number, that is,
(3) $q_0 + q_1e + \cdots + q_ne^n = 0$ for $n \geq 1$ and integers $q_0 \neq 0, q_1, \ldots, q_n$, and the greatest common divisor of $q_0, q_1, \ldots, q_n$, is equal to 1.

6.6. An Application – the Transcendence of e and π

We consider now the polynomial $f(x) = x^{p-1}(x-1)^p \ldots (x-n)^p$ with p a sufficiently large prime number, and we consider $I(z_1)$ with respect to this polynomial. Let

$$J = q_0 I(0) + q_1 I(1) + \cdots + q_n I(n).$$

From (1) and (3) we get that

$$J = -\sum_{j=0}^{m}\sum_{k=0}^{n} q_k f^{(j)}(k),$$

where $m = (n+1)p - 1$ since $(q_0 + q_1 e + \cdots + q_n e^n)(\sum_{j=0}^{m} F^{(j)}(0)) = 0$.

Now, $f^{(j)}(k) = 0$ if $j < p, k > 0$, and if $j < p-1$ then $k = 0$, and hence $f^{(j)}(k)$ is an integer that is divisible by $p!$ for all j, k except for $j = p-1, k = 0$. Further, $f^{(p-1)}(0) = (p-1)!(-1)^{np}(n!)^p$, and hence, if $p > n$, then $f^{(p-1)}(0)$ is an integer divisible by $(p-1)!$ but not by $p!$.

It follows that J is a nonzero integer that is divisible by $(p-1)!$ if $p > |q_0|$ and $p > n$. So let $p > n, p > |q_0|$, so that $|J| \geq (p-1)!$.

Now, $|f|(k) \leq (2n)^m$. Together with (2) we then get that

$$|J| \leq |q_1| e |f|(1) + \cdots + |q_n| n e^n |f|(n) \leq c^p$$

for a number c independent of p. It follows that

$$(p-1)! \leq |J| \leq c^p,$$

that is,

$$1 \leq \frac{|J|}{(p-1)!} \leq c \frac{c^{p-1}}{(p-1)!}.$$

This gives a contradiction, since $\frac{c^{p-1}}{(p-1)!} \to 0$ as $p \to \infty$. Therefore, e is transcendental. ∎

We now move on to the transcendence of π. We first need the following lemma.

Lemma 6.6.5
Suppose $\alpha \in \mathbb{C}$ is an algebraic number and $f(x) = a_n x^n + \cdots + a_0, n \geq 1, a_n \neq 0$, and all $a_i \in \mathbb{Z}$ ($f(x) \in \mathbb{Z}[x]$) with $f(\alpha) = 0$. Then $a_n \alpha$ is an algebraic integer.

Proof

$$a_n^{n-1} f(x) = a_n^n x^n + a_n^{n-1} a_{n-1} x^{n-1} + \cdots + a_n^{n-1} a_0$$
$$= (a_n x)^n + a_{n-1}(a_n x)^{n-1} + \cdots + a_n^{n-1} a_0$$
$$= g(a_n x) = g(y) \in \mathbb{Z}[y]$$

where $y = a_n x$ and $g(y)$ is monic. Then $g(a_n \alpha) = 0$, and hence $a_n \alpha$ is an algebraic integer. ∎

Theorem 6.6.2
π is a transcendental number, that is, transcendental over $\mathbb{Q}$.

Proof
Assume that π is an algebraic number. Then $\theta = i\pi$ is also algebraic. Let $\theta_1 = \theta, \theta_2, \ldots, \theta_d$ be the conjugates of θ. Suppose

$$p(x) = q_0 + q_1 x + \cdots + q_d x^d \in \mathbb{Z}[x], q_d > 0, \text{ and } \gcd(q_0, \ldots, q_d) = 1$$

is the entire minimal polynomial of θ over $\mathbb{Q}$. Then $\theta_1 = \theta, \theta_2, \ldots, \theta_d$ are the zeros of this polynomial. Let $t = q_d$. Then from Lemma 6.6.5, $t\theta_i$ is an algebraic integer for all i. From $e^{i\pi} + 1 = 0$ and from $\theta_1 = i\pi$ we get that

$$(1 + e^{\theta_1})(1 + e^{\theta_2}) \ldots (1 + e^{\theta_d}) = 0.$$

The product on the left side can be written as a sum of 2^d terms e^ϕ, where $\phi = \epsilon_1 \theta_1 + \cdots + \epsilon_d \theta_d$, $\epsilon_j = 0$ or 1. Let n be the number of terms $\epsilon_1 \theta_1 + \cdots + \epsilon_d \theta_d$ that are nonzero. Call these $\alpha_1, \ldots, \alpha_n$. We then have an equation

$$(4) q + e^{\alpha_1} + \cdots + e^{\alpha_n} = 0$$

with $q = 2^d - n > 0$. Recall that all $t\alpha_i$, are algebraic integers and we consider the polynomial

$$f(x) = t^{np} x^{p-1} (x - \alpha_1)^p \ldots (x - \alpha_n)^p$$

with p a sufficiently large prime integer. We have $f(x) \in \mathbb{R}[x]$, since the α_i are algebraic numbers and the elementary symmetric polynomials in $\alpha_1, \ldots, \alpha_n$ are rational numbers.

Let $I(z_1)$ be defined as in the proof of Theorem 6.6.1, and now let

$$J = I(\alpha_1) + \cdots + I(\alpha_n).$$

From (1) in the proof of Theorem 6.6.1 and (4) we get

$$J = -q \sum_{j=0}^{m} f^{(j)}(0) - \sum_{j=0}^{m} \sum_{k=1}^{n} f^{(j)}(\alpha_k),$$

with $m = (n+1)p - 1$.

Now, $\sum_{k=1}^{n} f^{(j)}(\alpha_k)$ is a symmetric polynomial in $t\alpha_1, \ldots, t\alpha_n$ with integer coefficients since the $t\alpha_i$ are algebraic integers. It follows from the main theorem on symmetric polynomials that $\sum_{j=0}^{m} \sum_{k=1}^{n} f^{(j)}(\alpha_k)$ is an integer. Further, $f^{(j)}(\alpha_k) = 0$ for $j < p$. Hence $\sum_{j=0}^{m} \sum_{k=1}^{n} f^{(j)}(\alpha_k)$ is an integer divisible by $p!$.

Now, $f^{(j)}(0)$ is an integer divisible by $p!$ if $j \neq p - 1$, and $f^{(p-1)}(0) = (p-1)!(-t)^{np}(\alpha_1 \ldots \alpha_n)^p$ is an integer divisible by $(p-1)!$ but not divisible by $p!$ if p is sufficiently large. In particular, this is true if $p > |t^n(\alpha_1 \ldots \alpha_n)|$ and also $p > q$.

From (2) in the proof of Theorem 6.6.1 we get that

$$|J| \leq |\alpha_1|e^{|\alpha_1|}|f|(|\alpha_1|) + \cdots + |\alpha_n|e^{|\alpha_n|}|f|(|\alpha_n|) \leq c^p$$

for some number c independent of p.

As in the proof of Theorem 6.6.1, this gives us

$$(p-1)! \leq |J| \leq c^p,$$

that is,

$$1 \leq \frac{|J|}{(p-1)!} \leq c \frac{c^{p-1}}{(p-1)!}.$$

This as before gives a contradiction, since $\frac{c^{p-1}}{(p-1)!} \to 0$ as $p \to \infty$. Therefore, π is transcendental. ∎

6.7 The Fundamental Theorem of Symmetric Polynomials

In our last proof of the Fundamental Theorem of Algebra we used the fact that any symmetric polynomial in n indeterminates is a polynomial in the elementary symmetric polynomials in these indeterminates. We used this important result, called the fundamental theorem of symmetric polynomials, again in our proofs of the transcendence of e and π. In this section we give a proof of this theorem.

Let R be an integral domain with $x_1, \ldots, x_n$ (independent) indeterminates over R and let $R[x_1, \ldots, x_n]$ be the polynomial ring in these indeterminates. Any polynomial $f(x_1, \ldots, x_n) \in R[x_1, \ldots, x_n]$ is composed of a sum of **pieces** of the form $ax_1^{i_1} \ldots x_n^{i_n}$ with $a \in R$. We first put an order on these pieces of a polynomial.

The piece $ax_1^{i_1} \ldots x_n^{i_n}$ with $a \neq 0$ is called **higher** than the piece $bx_1^{j_1} \ldots x_n^{j_n}$ with $b \neq 0$ if the first one of the differences

$$i_1 - j_1, i_2 - j_2, \ldots, i_n - j_n$$

that differs from zero is in fact positive. The highest piece of a polynomial $f(x_1, \ldots, x_n)$ is denoted by $HG(f)$.

Lemma 6.7.1
For $f(x_1, \ldots, x_n), g(x_1, \ldots, x_n) \in R[x_1, \ldots, x_n]$ we have $HG(fg) = HG(f) HG(g)$.

Proof

We use an induction on n, the number of indeterminates. It is clearly true for $n = 1$, and now assume that the statement holds for all polynomials in k indeterminates with $k < n$ and $n \geq 2$. Order the polynomials via exponents on the first indeterminate x_1 so that

$$f(x_1, \ldots, x_n) = x_1^r \phi_r(x_2, \ldots, x_n) + x_1^{r-1}\phi_{r-1}(x_2, \ldots, x_n)$$
$$+ \cdots + \phi_0(x_2, \ldots, x_n),$$

$$g(x_1, \ldots, x_n) = x_1^s \psi_s(x_2, \ldots, x_n) + x_1^{s-1}\psi_{s-1}(x_2, \ldots, x_n)$$
$$+ \cdots + \psi_0(x_2, \ldots, x_n).$$

Then $HG(fg) = x_1^{r+s} HG(\phi_r \psi_s)$. By the inductive hypothesis $HG(\phi_r \psi_s) = HG(\phi_r)HG(\psi_s)$. Hence

$$HG(fg) = x_1^{r+s} HG(\phi_r) HG(\psi_s) = (x_1^r HG(\phi_r))(x_1^s HG(\psi_s)) = HG(f)HG(g). \blacksquare$$

Recall (see Section 6.4) that the elementary symmetric polynomials in n indeterminates $x_1, \ldots, x_n$ are

$$s_1 = x_1 + x_2 + \cdots + x_n,$$
$$s_2 = x_1 x_2 + x_1 x_3 + \cdots + x_{n-1} x_n,$$
$$s_3 = x_1 x_2 x_3 + x_1 x_2 x_4 + \cdots + x_{n-2} x_{n-1} x_n,$$
$$\vdots$$
$$s_n = x_1 \ldots x_n.$$

These were found by forming the polynomial $p(x, x_1, \ldots, x_n) = (x - x_1) \cdots (x - x_n)$. The ith elementary symmetric polynomial s_i in $x_1, \ldots, x_n$ is then $(-1)^i a_i$, where a_i is the coefficient of x^{n-i} in $p(x, x_1, \ldots, x_n)$.

In general,

$$s_k = \sum_{i_1 < i_2 < \cdots < i_k, 1 \leq k \leq n} x_{i_1} x_{i_2} \ldots x_{i_k},$$

where the sum is taken over all the $\binom{n}{k}$ different systems of indices $i_1, \ldots, i_k$ with $i_1 < i_2 < \cdots < i_k$.

Further, a polynomial $s(x_1, \ldots, x_n)$ is a **symmetric polynomial** if $s(x_1, \ldots, x_n)$ is unchanged by any permutation σ of $\{x_1, \ldots, x_n\}$, that is, $s(x_1, \ldots, x_n) = s(\sigma(x_1), \ldots, \sigma(x_n))$.

Lemma 6.7.2
In the highest piece $ax_1^{k_1} \ldots x_n^{k_n}, a \neq 0$, of a symmetric polynomial $s(x_1, \ldots, x_n)$ we have $k_1 \geq k_2 \geq \cdots \geq k_n$.

6.7. The Fundamental Theorem of Symmetric Polynomials

Proof
Assume that $k_i < k_j$ for some $i < j$. As a symmetric polynomial, $s(x_1, \ldots, x_n)$ also must then contain the piece $ax_1^{k_1} \ldots x_i^{k_j} \ldots x_j^{k_i} \ldots x_n^{k_n}$, which is higher than $ax_1^{k_1} \ldots x_i^{k_i} \ldots x_j^{k_j} \ldots x_n^{k_n}$, giving a contradiction. ∎

Lemma 6.7.3
The product $s_1^{k_1-k_2} s_2^{k_2-k_3} \ldots s_{n-1}^{k_{n-1}-k_n} s_n^{k_n}$ with $k_1 \geq k_2 \geq \cdots \geq k_n$ has the highest piece $x_1^{k_1} x_2^{k_2} \ldots x_n^{k_n}$.

Proof
From the definition of the elementary symmetric polynomials we have that
$$HG(s_k^t) = (x_1 x_2 \ldots x_k)^t, \quad 1 \leq k \leq n, t \geq 1.$$
From Lemma 6.7.1,
$$HG(s_1^{k_1-k_2} s_2^{k_2-k_3} \ldots s_{n-1}^{k_{n-1}-k_n} s_n^{k_n}) = x_1^{k_1-k_2}(x_1 x_2)^{k_2-k_3} \ldots (x_1 \ldots x_{n-1})^{k_{n-1}-k_n}$$
$$(x_1 \ldots x_n)^{k_n} = x_1^{k_1} x_2^{k_2} \ldots x_n^{k_n}. \quad \blacksquare$$

Theorem 6.7.1
Let $s(x_1, \ldots, x_n) \in R[x_1, \ldots, x_n]$ be a symmetric polynomial. Then $s(x_1, \ldots, x_n)$ can be uniquely expressed as a polynomial $f(s_1, \ldots, s_n)$ in the elementary symmetric polynomials $s_1, \ldots, s_n$ with coefficients from R.

Proof
We prove the existence of the polynomial f by induction on the size of the highest pieces. If in the highest piece of a symmetric polynomial all exponents are zero, then it is constant, that is, an element of R and there is nothing to prove.

Now we assume that each symmetric polynomial with highest piece smaller than that of $s(x_1, \ldots, x_n)$ can be written as a polynomial in the elementary symmetric polynomials. Let $ax_1^{k_1} \ldots x_n^{k_n}$, $a \neq 0$, be the highest piece of $s(x_1, \ldots, x_n)$. Let
$$t(x_1, \ldots, x_n) = s(x_1, \ldots, x_n) - as_1^{k_1-k_2} \ldots s_{n-1}^{k_{n-1}-k_n} s_n^{k_n}.$$
Clearly, $t(x_1, \ldots, x_n)$ is another symmetric polynomial, and from Lemma 6.7.3 the highest piece of $t(x_1, \ldots, x_n)$ is smaller than that of $s(x_1, \ldots, x_n)$. Therefore, $t(x_1, \ldots, x_n)$ and hence $s(x_1, \ldots, x_n) = t(x_1, \ldots, x_n) + as_1^{k_1-k_2} \ldots s_{n-1}^{k_{n-1}-k_n} s_n^{k_n}$ can be written as a polynomial in $s_1, \ldots, s_n$.

To prove the uniqueness of this expression assume that $s(x_1, \ldots, x_n) = f(s_1, \ldots, s_n) = g(s_1, \ldots, s_n)$. Then $f(s_1, \ldots, s_n) - g(s_1, \ldots, s_n) = h(s_1, \ldots, s_n) = \phi(x_1, \ldots, x_n)$ is the zero polynomial in $x_1, \ldots, x_n$. Hence, if we write $h(s_1, \ldots, s_n)$ as a sum of products of powers of the $s_1, \ldots, s_n$, all coefficients disappear because two different products of powers in the $s_1, \ldots, s_n$ have different highest pieces. This follows from previous set of lemmas. Therefore, f and g are the same, proving the theorem. ∎

Exercises

6.1. Let $p(x) = x^2 + x + 1$.
   (a) Show that $p(x)$ is irreducible over $\mathbb{Q}$.
   (b) Let α be a root of $p(x)$. Let $g = 2 + \alpha$, $h = 1 - \alpha \in \mathbb{Q}(\alpha)$. Find $g + h, gh$, and h^{-1}.
   (c) Solve the linear equation $2g + h = 5$ in the field $\mathbb{Q}(\alpha)$.

6.2. Assume that α, β are both algebraic over F and suppose $irr(\alpha, F) = irr(\beta, F)$. Verify that the map given in the proof of Lemma 6.1.4 is indeed an F-isomorphism.

6.3. Let R be a commutative ring and $r \in R$. Let $(r) = \{r_1 r; r_1 \in R\}$. Verify that (r) is an ideal.

6.4. An ideal I is a **prime ideal** if $I \neq R$ and whenever $r_1 r_2 \in I$ either $r_1 \in I$ or $r_2 \in I$.
   (a) Show that $m\mathbb{Z}$ is a prime ideal in $\mathbb{Z}$ if and only if m is a prime.
   (b) For any commutative ring R with identity show that if I is a prime ideal, then the factor ring R/I is an integral domain.

6.5. If F is a field, show that the only ideals in F are the whole field and (0).

6.6. Let $I \subset R$ be an ideal and $r \in R$, $r \notin I$. Verify that $(r, I) = \{r_1 r + i_1; r_1 \in R, i_1 \in I\}$ is an ideal.

6.7. Let $F \subset E \subset F'$. If $\{\alpha_1, \ldots, \alpha_n\}$ is a basis for E over F and $\{\beta_1, \ldots, \beta_m\}$ is a basis for F' over E, then the mn products $\{\alpha_i \beta_j\}$ are a basis for F' over F.
   Hint: To show that $\{\alpha_i \beta_j\}$ is a basis for F' over F we must show that they (1) span F' over F and (2) that they are independent.
   To show (1): Let $g \in F'$. Then $g = e_1 \beta_1 + \cdots + e_m \beta_m$ with $e_i \in E$, since $\{\beta_1, \ldots, \beta_m\}$ is a basis of F' over E. Now use linear expansion for each e_i in terms of $\{\alpha_1, \ldots, \alpha_n\}$ and collect terms.
   To show (2): Suppose $f_1 \alpha_1 \beta_1 + \cdots + f_{mn} \alpha_n \beta_m = 0$. Combine terms in each β_i first and then use the independence of the $\{\beta_i\}$. Then use the independence of the $\{\alpha_j\}$ in each coefficient.

6.8. Let K be an algebraic extension of E and E an algebraic extension of F. Then K is an algebraic extension of F.

6.9. Let F' be a finite extension of F. Let $\alpha \in F'$ with $n = \deg irr(\alpha, F)$. Then n divides $|F' : F|$.

6.10. Let $F \subset D$, where F is a field and D is an integral domain with the same multiplicative identity. D is then a vector space over F. If D is finite-dimensional over F then D is a field.

Exercises

6.11. Write down the first five elementary symmetric polynomials.

6.12. Let D_3 be the group of symmetries of an equilateral triangle.
   (1) Show that $|D_3| = 6$.
   (2) Write down a multiplication table and a presentation for D_3.
   (3) Show that $S_3 \cong D_3$.

6.13. Let D_4 be the group of symmetries of a square. Show that $|D_4| = 8$ and find a presentation for D_4. Show that $D_4 \not\cong S_4$. In general, what do you think a presentation for the group of symmetries of a regular n-gon is?

CHAPTER 7

Galois Theory

7.1 Galois Theory Overview

In the last chapter we gave an algebraic proof of the Fundamental Theorem of Algebra. This depended on the facts that we could always construct a splitting field for a given polynomial, that odd degree real polynomials have real roots, and that complex numbers always have squareroots, implying that any quadratic polynomial is solvable in $\mathbb{C}$. In this chapter we give a proof suggested by the last proof but involving the more general ideas of Galois theory.

Galois theory is that branch of mathematics that deals with the interplay of the algebraic theory of fields, the theory of equations and finite group theory. Much of the foundation for Galois theory, involving algebraic extensions of fields, was introduced in the last chapter.

This theory was introduced by Evariste Galois about 1830 in his study of the insolvability by radicals of quintic (degree 5) polynomials, a result proved somewhat earlier by Ruffini and independently by Abel. Galois was the first to see the close connection between field extensions and permutation groups. In doing so he initiated the study of finite groups. He was the first to use the term group, as an abstract concept although his definition was really just for a closed set of permutations.

The method Galois developed not only facilitated the proof of the insolvability of the quintic and higher powers but led to other applications and to a much larger theory as well. In this chapter, however, we will only examine those parts of the theory relevant to the Fundamental Theorem of Algebra.

7.2. Some Results From Finite Group Theory

The main idea of Galois theory is to associate to certain special types of algebraic field extensions called **Galois extensions** a group called the **Galois group**. The properties of the field extension will be reflected in the properties of the group, which are somewhat easier to examine. Thus, for example, solvability by radicals can be translated into a group property called solvability of groups. Showing that for every degree five or greater there exists a field extension whose Galois group does not have this property proves that there cannot be a general formula for solvability by radicals.

The tie-in to the theory of equations is as follows: If $f(x) = 0$ is a polynomial equation over some field F, we can form the splitting field K. This is usually a Galois extension, and therefore has a Galois group called the **Galois group of the equation**. As before, properties of this group will reflect properties of this equation.

Galois theory depends in part on the theory of finite groups, and so in the next section we review some of the basic necessary results from this theory. In Section 7.3 we introduce the properties of normality and separability of field extensions that define Galois extensions and in the subsequent section develop the Galois group and its construction. We next summarize all the results in the **fundamental theorem of Galois theory** which describes the interplay between the Galois group and Galois extensions. With all the machinery in place, in Section 7.6 we give our fourth proof of the Fundamental Theorem of Algebra. Finally, we close the chapter by giving two additional applications of Galois theory. The first is a sketch of the proof of the insolvability of the quintic, while the second is a discussion of certain geometric ruler and compass constructions and their algebraic interpretations.

Since our aim is to arrive rather quickly at the main results of Galois theory and then give our fourth proof of the Fundamental Theorem of Algebra many of the more difficult proofs along the way will be omitted.

7.2 Some Results From Finite Group Theory

In this section we review some basic results from finite group theory. Recall (see Section 6.4) that a **group** G is a set with one binary operation (which we will denote by multiplication) such that

(1) The operation is associative.
(2) There exists an identity for this operation.
(3) Each $g \in G$ has an inverse for this operation.

If, in addition, the operation is commutative, the group G is called an **abelian group**. The **order** of G is the number of elements in G, denoted by $|G|$. If $|G| < \infty$, G is a **finite group**. $H \subset G$ is a **subgroup** if $H \neq \emptyset$ and a group under the same operation as G. Equivalently, H is a subgroup if $H \neq \emptyset$ and H is closed under the operation and inverses.

As we indicated in Section 6.4 groups most often arise from invertible mappings of a set onto itself. Such mappings are called **permutations**. The group of all permutations on a set with n elements is called the **symmetric group** on n symbols, denoted by S_n.

As a variation of permutation groups we obtain automorphism groups. Recall that if R and S are algebraic structures (groups, rings, fields, vector spaces etc.) then a mapping $F : R \to S$ is a **homomorphism** if it preserves the algebraic operations. By this we mean:

(1) $f(r_1 r_2) = f(r_1)f(r_2)$ if R, S are groups.
(2) $f(r_1 + r_2) = f(r_1) + f(r_2)$ and $f(r_1 r_2) = f(r_1)f(r_2)$ if R, S are rings or fields.
(3) $f(r_1 + r_2) = f(r_1) + f(r_2)$ and $f(cr_1) = cf(r_1)$ with c an element of the scalar field if R, S are vector spaces

A homomorphism $f : R \to S$ that is also a bijection is an **isomorphism**.

Definition 7.2.1
If R is an algebraic structure then an **automorphism** of R is an isomorphism $\sigma : R \to R$. We let **Aut(R)** denote the set of automorphisms on R.

Example 7.2.1
Let G be a cyclic group of order n. Suppose g is a generator so that $G = \{1, g, g^2, \ldots, g^{n-1}\}$. Recall that if $(k, n) = 1$, then g^k is also a generator, so the mapping

$$\sigma : g \to g^k$$

will define an automorphism of G by making this map a homomorphism. Further, for any automorphism σ the generator g is mapped to g_1 where g_1 is another generator. Therefore,

$$\text{Aut}(G) = \{\sigma; \sigma : g \to g^k \text{ with } (n, k) = 1.\}$$

It follows that $|\text{Aut}(G)| = \phi(n)$, the number of positive integers less than n and relatively prime to n. □

Example 7.2.2
Consider the complex numbers $\mathbb{C}$. Let σ be an automorphism of $\mathbb{C}$. For all such automorphisms it follows that $\sigma(0) = 0, \sigma(1) = 1, \sigma(-1) = -1$. Therefore, $\sigma(i^2) = (\sigma(i))^2 = -1$ and hence $\sigma(i) = \pm i$.

7.2. Some Results From Finite Group Theory

Since $1, i$ form a basis for $\mathbb{C}$ over $\mathbb{R}$, the images of $1, i$ will completely determine an automorphism. Hence there are precisely two automorphisms of $\mathbb{C}$, namely those given by

$$\sigma_1 : 1 \to 1, i \to i, \text{ the identity automorphism}$$
$$\sigma_2 : 1 \to 1, i \to -i.$$

□

EXAMPLE 7.2.3
Consider the finite field $\mathbb{Z}_p$, where p is a prime. Then $\mathbb{Z}_p = \{0, 1, \ldots, p-1\}$ with the arithmetic operations done modulo p. Let σ be an automorphism. Since $\sigma(1) = 1$, it follows that $\sigma(n) = n$ for all integer multiples of 1. Hence $\sigma(x) = x$ for all $x \in \mathbb{Z}_p$. Therefore, the only automorphism on $\mathbb{Z}_p$, is the identity and the automorphism group is trivial. □

Theorem 7.2.1
*For any algebraic structure R the set $\mathrm{Aut}(R)$ forms a group under composition called the **automorphism group** of R. If S is any substructure of R, then those automorphisms $\sigma \in \mathrm{Aut}(R)$ that **fix** S, that is, $\sigma(s) = s$ for all $s \in S$ form a subgroup.*

EXAMPLE 7.2.4
Let p be a prime and G a cyclic group of order p. We show that $\mathrm{Aut}(G)$ is also cyclic and of order $p - 1$.

Let $\mathbb{Z}_p$ be the finite field of order p as in the last example. Its additive group is cyclic of order p and so isomorphic to G under the isomorphism $1 \to g$, where g is a generator of G. Since $\mathbb{Z}_p$ is a field, its nonzero elements form a group under multiplication. We will call this group U.

From Example 7.2.1, any automorphism of G is determined by the image of a generator

$$\sigma : g \to g^k, \text{ where } (k, p) = 1.$$

In terms of the isomorphism with the additive group of $\mathbb{Z}_p$ this is then given by

$$\sigma : 1 \to k, \text{ where } k \neq 0.$$

This automorphism then behaves just like multiplication by k in $\mathbb{Z}_p$, and so $\mathrm{Aut}(G)$ is isomorphic to the multiplicative group U of $\mathbb{Z}_p$. We show that U is cyclic.

Now, $|U| = p - 1$, so $y^{p-1} = 1$ for all $y \in U$. Let m be the maximal order of any element in U, so $m \leq p - 1$. If $y \in U$ has order q, then $q|m$ (see the exercises). Hence, if $y^q = 1$, then $y^m = 1$. It follows that $y^m = 1$ for all $y \in U$. Therefore, each $y \in U$ is a root of the polynomial $P(x) = x^m - 1$. This is a polynomial over a field, so it can have at most m roots. However, it has at least $p - 1$ roots so $m \geq p - 1$. Therefore, $m = p - 1$, and U has an element of order $p - 1$ and therefore is cyclic. □

We now discuss the subgroup structure of a finite group. If G is a group and H a subgroup of G recall that a **left coset** of H in G is a set $gH = \{gh; h \in H\}$ for some $g \in G$. Similarly, a **right coset** is $Hg = \{hg; h \in H\}$ for some $g \in G$. It can be shown that the set of left (right) cosets partition G, and each has the same size as the subgroup H. It follows that there are the same number of left and right cosets. This number is called the **index** of H in G, denoted by $|G : H|$. Applying these ideas to finite groups, we get Lagrange's theorem.

Theorem 7.2.2
(*Lagrange's Theorem*) *Let G be a finite group and H be a subgroup. Then $|G| = |G{:}H||H|$. In particular, both the order of a subgroup and the index of a subgroup divide the order of a finite group.*

One interesting historical note. Lagrange died about twenty years before the word "group" was ever used. What Lagrange actually proved was a result, equivalent to what is now called Lagrange's theorem, on closed sets of permutations.

EXAMPLE 7.2.5
We show that every finite group of prime order is cyclic.

Let p be a prime and G a finite group of order p. Let $x \in G, x \neq 1$, and let $H = <x>$ the cyclic subgroup generated by x. From Lagrange's theorem, the order of H divides the order of G. Hence $|H| = 1$ or p, since p is a prime. If $|H| = 1$, then $x = 1$, contradicting that $x \neq 1$. Therefore, $|H| = p$ and $H = G$ and G is cyclic.

The above argument also shows that a finite group of prime order can have no nontrivial proper subgroups. □

We now examine certain special types of subgroups called normal subgroups. From these we can build factor groups. These stand in the same relation to group theory as ideals and factor rings are to ring theory (see Section 6.2).

Definition 7.2.2
If G is a group, $H \subset G$ a subgroup, and $g \in G$, then $g^{-1}Hg$ forms a subgroup, called a **conjugate** of H. g **normalizes** H if $g^{-1}Hg = H$. The set of elements that normalize H is called the **normalizer** of H in G denoted by $N_G(H)$. H is a **normal subgroup** denoted by $H \triangleleft G$, if every $g \in G$ normalizes H, that is, $g^{-1}Hg = H$ for all $g \in G$.

Notice that if $H \triangleleft G$ then $g^{-1}Hg = H$. This implies that $Hg = gH$, or in other words the left coset gH is the same as the right coset Hg. We summarize some of the properties of conjugation, normalizer and normal subgroups in the following two lemmas.

7.2. Some Results From Finite Group Theory

Lemma 7.2.1
The following are equivalent:

(1) $H \triangleleft G$.
(2) *The only conjugate of H in G is H.*
(3) *Each left coset of H is also a right coset.*
(4) $N_G(H) = G$.

Lemma 7.2.2
Let $H \subset G$ be a subgroup. Then:

(1) *Any conjugate of H is isomorphic to H.*
(2) $N_G(H)$ *is a subgroup of G and* $H \triangleleft N_G(H)$.
(3) $|G:N_G(H)|$ *is the number of distinct conjugates of H in G.*

EXAMPLE 7.2.6
In an abelian group G every subgroup H is normal, since $g^{-1}Hg = H(g^{-1}g) = H$.

At the other end of the spectrum are **simple groups**. These are groups that have no proper nontrivial normal subgroups. For example, a cyclic group of prime order is simple, since it has no proper nontrivial subgroups whatsoever.

Simple groups are really the building blocks for finite groups, and there is a vast theory of finite simple groups. Much of the work has gone into the proof of what is called the **classification theorem for finite simple groups**. This theorem gives a complete "listing" of all the finite simple groups. It turns out that any finite simple group is either a member of one of a finite number of infinite families of groups that are well described or is one of a finite number of isolated examples called sporadic groups. The proof of this theorem, at this point in time, takes over twelve thousand published pages. □

EXAMPLE 7.2.7
Suppose $|G : H| = 2$. We show that $H \triangleleft G$.

Let $g \notin H$, then there are two left cosets namely $H = 1H$ and gH. These partition G, so that

$$G = H \cup gH$$

with this being a disjoint union. A similar statement is true for right cosets, so that

$$G = H \cup Hg.$$

Since these are disjoint unions, it follows that $gH = Hg$, or $g^{-1}Hg = H$.

As a specific example of this, consider the symmetric group on three symbols S_3. In Example 6.4.1 we saw that $|S_3| = 6$ and that there is an element of order 3. By Lagrange's theorem the cyclic subgroup generated by this element has index two and is thus normal. This subgroup is called A_3 – the alternating group on three symbols. In a similar way, for each n, there is a subgroup of S_n of index two (not cyclic for $n > 3$ however) called A_n the alternating group on n symbols. A_n consists of the even permutations (see Exercise 7.18). □

Normal subgroups allow us to form **factor groups**. Suppose $H \triangleleft G$. Then let G/H = set of cosets of H in G. Since each left coset is also a right coset, we don't need to specify left or right. On G/H define

$$(g_1 H)(g_2 H) = g_1 g_2 H.$$

Since $Hg_2 = g_2 H$ and $HH = H$, this operation is well-defined and allows us to obtain a group.

Lemma 7.2.3
*If $H \triangleleft G$, then G/H forms a group called the **factor group**, or **quotient group**, of G modulo H.*

Nomal subgroups and factor groups are closely tied to homomorphisms.

Definition 7.2.3
If $f : G \to H$ is a homomorphism, then the **kernel** of f, denoted by **ker** f is the set $\{g \in G; f(g) = 1\}$. The **image** of f, denoted by **Im** f is the set $\{h \in H; f(g) = h \text{ for some } g \in G\}$.

The notions of kernel, image, normal subgroup, and factor group are all tied together by the following theorem, which is called the first isomorphism theorem.

Theorem 7.2.3
(First Isomorphism Theorem) (1) If $f:G \to H$, is a homomorphism then ker $f \triangleleft G$, Im f is a subgroup of H, and

$$G/\ker f \cong \operatorname{Im} f$$

*(Note: If f is onto, then $G/\ker f \cong H$, and H is called a **homomorphic image** of G).*

(2) If $H \triangleleft G$, then there exists a homomorphism $f:G \to G/H$ such that ker $f = H$ and Im $f = G/H$.

Finally, for our proof of the Fundamental Theorem of Algebra we need some facts about p-groups and p-subgroups when p is a prime.

Definition 7.2.4
If p is a prime, then a p-**group** is a group G where every element has order a power of p. If G is finite, this implies that $|G| = p^n$ for some n.

Lemma 7.2.4
If G is a finite p-group of order p^n, then G has a subgroup of order p^{n-1} and hence of index p.

Notice that this lemma can be considered as a converse of Lagrange's theorem for *p*-groups. In general, Lagrange's theorem says that the order of a subgroup divides the order of the group. However, it does not imply that for any divisor of the order of the group there is a subgroup of that order. As a consequence of Lemma 7.2.4 this is true, however, for all finite *p*-groups.

Corollary 7.2.1
If G is a finite p-group, then for each divisor of $|G|$ there is a subgroup of G of that order.

We also need the Sylow theorem which provides a partial converse for Lagrange's theorem.

Definition 7.2.5
Suppose G is a finite group with $|G| = p^m \alpha$, with p a prime and with $(p, \alpha) = 1$. Then a *p*-**Sylow subgroup** is a subgroup of order p^m (a maximal *p*-subgroup of G).

Theorem 7.2.4
(Sylow Theorem) Let G be a finite group of order $p^m \alpha$ with p a prime and with $(p, \alpha) = 1$. Then:

(1) *G has a p-Sylow subgroup.*
(2) *All p-Sylow subgroups in G are conjugate.*
(3) *Any p-subgroup of G is contained in a p-Sylow subgroup.*
(4) *The number of p-Sylow subgroups is congruent to 1 modulo p and divides $|G|$ and hence divides α.*

EXAMPLE 7.2.8
We show that no group of order 12 is simple.

As a consequence of part (2) of Theorem 7.2.4, if there is only one *p*-Sylow subgroup in G it must be normal. Now, $|G| = 12 = 2^2 3$. We show that there is either only one 3-Sylow subgroup or only one 2-Sylow subgroup.

The number of 3-Sylow subgroups is of the form $1 + 3k$ and divides 4, so there exists either one or four 3-Sylow subgroups. If there is only one it is normal so suppose that there are four. Each of these has order 3, and therefore they intersect trivially. It follows that these four subgroups cover eight elements in G, not including the identity.

Each 2-Sylow subgroup has four elements. There is only trivial intersection between the elements in the 2-Sylow subgroups and those in the

3-Sylow subgroups. If there were more than one 2-Sylow subgroup there would be more than three additonal elements outside the eight within the 3-Sylow subgroups. This is impossible since G contains only twelve elements. Therefore, there must be only one 2-Sylow subgroup and hence it must be normal. □

7.3 Galois Extensions

Galois theory deals with certain special types of finite algebraic extensions. In particular, we need two special properties – separability and normality. Normality is the simpler one so we discuss that first. For the remainder of the chapter all extensions are to be considered finite extensions.

Definition 7.3.1
K is a **normal extension** of a ground field F if K is a splitting field over F.

Several facts about normal extensions are crucial for us. These are given in the next theorem.

Theorem 7.3.1
Suppose K is a normal extension of F and suppose $F \subset E \subset K \subset \overline{F}$, where $\overline{F}$ is an algebraic closure of F. Then:

(1) *Any automorphism of $\overline{F}$ leaving F fixed maps K onto itself and is thus an automorphism of K leaving F fixed. Thus any isomorphism of K within $\overline{F}$ leaving F fixed is actually an automorphism.*
(2) *Every irreducible polynomial in $F[x]$ having a root in K splits in K.*
(3) *K is a normal extension of E.*

The other major property is separability. This concerns multiplicity of roots.

Definition 7.3.2
If α is a root of $f(x)$ then α has **multiplicity** $m \geq 1$ if $f(x) = (x - \alpha)^m g(x)$, where $g(\alpha) \neq 0$. If $m = 1$, then α is a **simple root** otherwise it is a **multiple root**.

Now, suppose K is a finite extension of F and $\alpha \in K$. Then α is **separable** over F if α is a simple root of $irr(\alpha, F)$. K is a **separable extension** if every $\alpha \in K$ is separable over F.

Thus in a separable extension of F, if $\alpha \notin F$, then α is not a multiple nonsimple root of its irreducible polynomial.

7.3. Galois Extensions

Although separability is an essential property for Galois extensions it will not play a major role in the Fundamental Theorem of Algebra since the ground fields we work with are extensions of $\mathbb{Q}$, $\mathbb{R}$, or $\mathbb{C}$. All these fields have characteristic zero and this forces any extension to be separable (see Section 6.6).

Definition 7.3.3
A field F has **characteristic** n if n is the least positive integer such that $(n)(1) = 0$ in F. We denote this by char $F = n$. If no such n exists, we say that F has **characteristic zero** denoted by char $F = 0$.

EXAMPLE 7.3.1
char $\mathbb{Q}$ = char $\mathbb{R}$ = char $\mathbb{C}$ = 0, and thus any extension of these has characteristic zero.

On the other hand, char $\mathbb{Z}_p = p$, and thus any extension of $\mathbb{Z}_p$ also has characteristic p. □

We give some simple facts about characteristic.

Lemma 7.3.1
The characteristic of a field is zero or a prime.

Proof
Suppose char $F = n \neq 0$. If n is composite then $n = mk$, with $m < n$ and $k < n$. Then $(n)(1) = (mk)(1) = ((m)(1))((k)(1)) = 0$. A field has no zero divisors, so either $(m)(1) = 0$ or $(k)(1) = 0$ contradicting the minimality of n. ∎

Lemma 7.3.2
If char $F = 0$, then F contains a subfield isomorphic to $\mathbb{Q}$. If char $F = p$, then F contains a subfield isomorphic to $\mathbb{Z}_p$. In particular, a field of characteristic zero must be infinite.

Proof
We show the prime case; the zero characteristic case is similar. Suppose char $F = p$. Since F is a field, it has an identity 1. Let $E = \{(n)(1); n \in \mathbb{Z}\}$. Since char $F = p$, $E = \{0, 1, (2)(1), \ldots, (p-1)(1)\}$. It is then straightforward to check that the map $(k)(1) \to k$ from E to $\mathbb{Z}_p$ is an isomorphism. ∎

The relevance of characteristic to separability is the following theorem.

Theorem 7.3.2
Any extension of a field of characteristic zero must be a separable extension.

In fact, any extension of a finite field is also separable so the only bad cases are infinite fields of characteristic p. For our purposes, what is important is that any extension of $\mathbb{Q}, \mathbb{R}$, or $\mathbb{C}$ is separable.

Separable extensions are essential to the interplay between field extensions and group theory because of the following two results, for the second of which we give the proof.

Theorem 7.3.3
If K is a finite separable extension of F, then the number of automorphisms of K fixing F is finite and equal to the degree $|K:F|$.

Theorem 7.3.4
(Primitive Element Theorem) If K is a finite separable extension of F, then K is a simple extension. That is, $K = F(\alpha)$ for some $\alpha \in K$.

Proof
Since K is a finite extension, $K = F(\alpha_1, \ldots, \alpha_n)$ for some elements $\alpha_1, \ldots, \alpha_n \in K$. By induction, it is enough to show that if $K = F(\alpha, \beta)$, then $K = F(\gamma)$ for some $\gamma \in K$.

Let $n = |K : F|$. From Theorem 7.3.3 there are then n automorphisms $\sigma_1, \ldots, \sigma_n$ of K fixing F. Form the polynomial

$$p(x) = \prod_{i \neq j}(\sigma_i(\alpha) + x\sigma_i(\beta) - \sigma_j(\alpha) - x\sigma_j(\beta)).$$

This is not the zero polynomial, so there exists a $c \in K$ with $p(c) \neq 0$. Then the elements $\sigma_i(\alpha + c\beta), i = 1, \ldots, n$, are distinct. It follows that $F(\alpha + c\beta)$ has degree at least n over F. But $F(\alpha + c\beta) \subset F(\alpha, \beta)$, and $F(\alpha, \beta)$ has degree n over F. Therefore, $F(\alpha, \beta) = F(\alpha + c\beta)$.

If $K = F(\alpha)$, α is called a **primitive element** of K over F. ∎

With these properties introduced we can define Galois extensions.

Definition 7.3.4
A **Galois extension** of F is a finite separable normal extension, that is, a finite separable splitting field over F.

Notice that if F has characteristic zero, then a Galois extension is just a finite extension splitting field over F.

Suppose that F is a field of characteristic zero, so that all extensions are separable, and suppose that $f(x) \in F[x]$ is an irreducible polynomial. If K is the splitting field for $f(x)$ over F, then K is a Galois extension of F. K is called the **Galois extension of F relative to** $f(x)$.

We close this section by summarizing what we know so far about Galois extensions.

Suppose $F \subset E \subset K \subset \overline{F}$ with K Galois over F and $\overline{F}$ an algebraic closure of F. Then:

(1) K is also Galois over E.
(2) The number of automorphisms of K fixing F is equal to the degree $|K : F|$.
(3) Any isomorphism of K within $\overline{F}$ fixing F is actually an automorphism of K.
(4) K is a simple extension of F.

7.4 Automorphisms and the Galois Group

We now introduce the Galois group and discuss the interplay between the group theory and the field extensions. We suppose that K is a finite extension of F. Recall from Section 7.2 that Aut(K) forms a group and that the elements of Aut(K) that fix F form a subgroup. If K is Galois over F, this subgroup is the Galois group.

Definition 7.4.1
Let K be a Galois extension of F. Then the group of automorphisms of K that fix F is called the **Galois group** of K over F, denoted by Gal(K/F). If H is a subgroup of Gal(K/F), we let K^H denote the elements of K fixed by H.

Since K is Galois over F, it is separable, and then from Theorem 7.3.3 we have:

Lemma 7.4.1
$|Gal(K/F)| = |K{:}F|$.

If E is an intermediate field, then K is also Galois over E. Those automorphisms in Gal(K/F) that also fix E form a subgroup. Thus Gal(K/E) is a subgroup of Gal(K/F). Conversely, if H is a subgroup of Gal(K/F), then K^H is an intermediate field and Gal(K/K^H) = H.

Lemma 7.4.2
Suppose $F \subset E \subset K$ with K Galois over F. Then:

(1) K is Galois over E and Gal(K/E) is a subgroup of Gal(K/F).
(2) If H is a subgroup of Gal(K/F), then $E = K^H$ is an intermediate field and Gal(K/E) = H.

Proof
Part (1) is clear. For part (2) we must show three things: that K^H is a field, that $F \subset K^H$, and that $\text{Gal}(K/K^H) = H$.

Since $H \subset \text{Gal}(K/F)$, every element of F is fixed by each element of H. Therefore, $F \subset K^H$ and hence K^H is not empty. To show that it forms a field we must show closure under the field operations.

Suppose $k_1, k_2 \in K^H$ and $\sigma \in H$. Then $\sigma(k_1) = k_1, \sigma(k_2) = k_2$ and so $\sigma(k_1 \pm k_2) = \sigma(k_1) \pm \sigma(k_2) = k_1 \pm k_2$. Therefore, $k_1 \pm k_2 \in K^H$. Similarly, $\sigma(k_1 k_2^{\pm 1}) = \sigma(k_1)(\sigma(k_2))^{\pm 1} = k_1 k_2^{\pm 1}$, for $k_2 \neq 0$. Hence $k_1 k_2^{\pm 1} \in K^H$ for $k_2 \neq 0$ and therefore, K^H is an intermediate field.

Finally, if $E = K^H$, then $\text{Gal}(K/E)$ consists of those automorphisms of K that fix E. But by definition E consists of those elements of K fixed by elements of H. Therefore, $\text{Gal}(K/E) = H$. ∎

Corollary 7.4.1
The map $\tau: H \to K^H$ from subgroups H of $\text{Gal}(K/F)$ to intermediate fields is a bijection.

From the previous results we have that if $F \subset E \subset K$ with K Galois over F, then K is Galois over E. The question naturally arises as to when E is also Galois over F. E is separable over F, so the question then becomes when is E a normal extension. This has the simple and elegant answer that E is a normal extension if and only if the corresponding subgroup of $\text{Gal}(K/F)$ is a normal subgroup.

Lemma 7.4.3
Suppose $F \subset E \subset K$ with K Galois over F and suppose $E = K^H$. Then E is Galois over F if and only if $H \triangleleft \text{Gal}(K/F)$. In this case

$$\text{Gal}(E/F) \cong G/H \cong \text{Gal}(K/F)/\text{Gal}(K/E).$$

The proofs of all these lemmas can be done directly but can also be done somewhat more easily by using the following theorem of Artin.

Theorem 7.4.1
(Artin) Let K be a field and G a finite group of automorphisms of K with $|G| = n$. Then K is a finite Galois extension of $F = K^G$ and its Galois group is G.

Proof
A separable polynomial is one with no multiple roots. If α is a root of a separable polynomial $f(x)$, then $\text{irr}(\alpha, F)$ divides $f(x)$, so α is separable over F.

Suppose $F = K^G$ and $\alpha \in K$. Let $g_1, \ldots, g_r$ be a maximal set of elements such that $g_1(\alpha), \ldots, g_r(\alpha)$ are all distinct. If $g \in G$, then $\{gg_1(\alpha), \ldots, gg_r(\alpha)\}$

7.4. Automorphisms and the Galois Group

is a permutation of the set $\{g_1(\alpha), \ldots, g_r(\alpha)\}$, since g is an automorphism and the set $\{g_1, \ldots, g_r\}$ is maximal. Therefore, α is a root of the polynomial

$$f(x) = \prod_{i=1}^{r}(x - g_i(\alpha)).$$

Further, $f(x)$ is fixed by any $g \in G$. Hence the coefficients of $f(x)$ lie in $F = K^G$. Since the $g_i(\alpha)$ are distinct, $f(x)$ is separable, and hence every $\alpha \in K$ is a root of a separable polynomial of degree $\leq n$ with coefficients in F. Therefore, K is a separable extension of F. Further, since $f(x)$ splits into linear factors K is a normal extension. Therefore, K is Galois over F.

Since K is separable over F, it is a simple extension, say $K = F(\gamma)$. But γ is a root of a polynomial of degree $\leq n$, and thus $|K : F| \leq n$. But $G \subset \text{Gal}(K/F)$, since G consists of automorphisms of K fixing F. Therefore, $|K : F| = |\text{Gal}(K/F)| \geq |G| = n$. It follows that $|K : F| = n$, and G must be the whole Galois group. ∎

The results we have just outlined are the main results in Galois theory. In the next section we summarize them and give examples. For now, we examine the Galois group of a polynomial. We suppose here and in the remainder of the section that F has characteristic zero, so that all extensions are separable.

Definition 7.4.2
Let $f(x)$ be an irreducible polynomial over F and let K be the splitting field for $f(x)$. Then K is Galois over F, and the corresponding Galois group is called the **Galois group of the polynomial**.

We examine this Galois group in more detail. Let K be any algebraic extension of F. Suppose $\alpha \in K$ with $irr(\alpha, F)$ its irreducible polynomial. Any other root $\bar{\alpha}$ of $irr(\alpha, F)$ is called a **conjugate** of α. Now, suppose K is Galois over F and $\sigma \in \text{Gal}(K/F)$. Since σ fixes F it also fixes $irr(\alpha, F) \in F[x]$, and hence $\sigma(\alpha)$ must be another root of $irr(\alpha, F)$. Therefore, any $\sigma \in \text{Gal}(K/F)$ maps elements to their conjugates.

Now, let $f(x) \in F[x]$ be irreducible, with roots $\alpha_1, \ldots, \alpha_n$ in K. If $\sigma \in \text{Gal}(K/F)$, σ fixes $f(x)$ and so maps this set of roots onto itself. Therefore, we have two facts: any $\sigma \in \text{Gal}(K/F)$ permutes the roots of irreducible polynomials, and conjugates of roots of irreducible polynomials are also roots. We also have the following.

Lemma 7.4.4
If K is Galois over F with $|K:F| = n$, then $\text{Gal}(K/F)$ is a subgroup of the symmetric group S_n.

Since $|\text{Gal}(K/F)| = n$ by Cayley's theorem it is a permutation group on itself and so a subgroup of S_n. What is important is how the permutations from $\text{Gal}(K/F)$ are obtained. Since K is a finite extension, $K = F(\alpha_1, \ldots, \alpha_r)$ for elements $\alpha_1, \ldots, \alpha_r \in K$. Those permutations that map $\alpha_1, \ldots, \alpha_r$ onto conjugates will lead to automorphisms and will give the elements of the Galois group. We illustrate this with an example.

EXAMPLE 7.4.1
Consider the splitting field of $x^4 + 1$ over $\mathbb{Q}$. In $\mathbb{C}$ this factors as $(x^2 + i)(x^2 - i)$, so this polynomial has four roots in $\mathbb{C}$ namely

$$\omega_1 = \frac{1+i}{\sqrt{2}} = \sqrt{i}, \quad \omega_2 = \frac{1-i}{\sqrt{2}} = \overline{\omega_1},$$

$$\omega_3 = \frac{-1+i}{\sqrt{2}} = i\omega_1, \quad \omega_4 = \frac{-1-i}{\sqrt{2}} = \overline{\omega_3}.$$

Notice that ω_1 is a primitive eighth root of unity, and $\omega_1^3 = \omega_3$, $\omega_1^5 = \omega_4$, $\omega_1^7 = \omega_2$, $\omega_1^8 = 1$.

Therefore, ω_1 is a primitive element of K over $\mathbb{Q}$, $\text{irr}(\omega_1, \mathbb{Q}) = x^4 + 1$, and $K = \mathbb{Q}(\omega_1)$. It follows that $|K : \mathbb{Q}| = 4$. Let us examine the Galois group.

Since ω_1 is a primitive root an automorphism σ is completely determined by its action on ω_1. There are exactly four elements in the Galois group and four possible conjugates of ω_1. Mapping ω_1 onto each will determine the four different automorphisms.

Suppose $\sigma_1(\omega_1) = \omega_1$. Then σ_1 fixes $\omega_1^3, \omega_1^5, \omega_1^7$, and thus σ_1 fixes K and is therefore the identity.

Suppose $\sigma_2(\omega_1) = \omega_2 = \omega_1^7$. Then $\sigma_2(\omega_2) = \sigma_2(\omega_1^7) = \omega_1^{49} = \omega_1$. Similarly, $\sigma_2(\omega_3) = \omega_4$, $\sigma_2(\omega_4) = \omega_3$. Therefore, the automorphism σ_2 is given by the permutation

$$\omega_1 \to \omega_2, \omega_2 \to \omega_1, \omega_3 \to \omega_4, \omega_4 \to \omega_3.$$

The same type of analysis gives that the two remaining automorphisms are given by the permutations

$$\sigma_3 : \omega_1 \to \omega_3, \omega_2 \to \omega_4, \omega_3 \to \omega_1, \omega_4 \to \omega_2.$$

$$\sigma_4 : \omega_1 \to \omega_4, \omega_2 \to \omega_3, \omega_3 \to \omega_2, \omega_4 \to \omega_1.$$

Checking, we see that each of these permutations has order 2. The group $G = \text{Gal}(K/\mathbb{Q})$ then has the presentation

$$G = <\sigma_2, \sigma_3; \sigma_2^2 = \sigma_3^2 = (\sigma_2\sigma_3)^2 = 1>.$$

It is easy to see that this group is then abelian and equal to $\mathbb{Z}_2 \times \mathbb{Z}_2$, the direct product of two finite cyclic groups of order 2. This group is usually referred to as the Klein 4-group.

7.5. The Fundamental Theorem of Galois Theory

Another way to see the structure of Gal($K/\mathbb{Q}$) is as follows. Notice that since $\omega_1 = \sqrt{i} \in K$, then $i \in K$. It follows that $1 + i \in K$, and hence $\frac{1+i}{\omega_1} = \sqrt{2} \in K$. Therefore, $\mathbb{Q}(i, \sqrt{2}) \subset K$. Since $|\mathbb{Q}(i, \sqrt{2}) : \mathbb{Q}| = 4$, it follows that $K = \mathbb{Q}(i, \sqrt{2})$. Now, $irr(i, \mathbb{Q}) = x^2 + 1$, so the conjugates of i are $\pm i$, while $irr(\sqrt{2}, \mathbb{Q}) = x^2 - 2$, so the conjugates of $\sqrt{2}$ are $\pm\sqrt{2}$. Therefore, the four possible automorphisms in Gal($K/\mathbb{Q}$) are given by:

$\sigma_1 : i \to i, \sqrt{2} \to \sqrt{2}$, the identity automorphism

$\sigma_2 : i \to -i, \sqrt{2} \to \sqrt{2}$;

$\sigma_3 : i \to i, \sqrt{2} \to -\sqrt{2}$;

$\sigma_4 : i \to -i, \sqrt{2} \to -\sqrt{2}$.

It is very clear here that each automorphism has order 2. □

7.5 The Fundamental Theorem of Galois Theory

We now summarize the results of the last two sections into one theorem called the fundamental theorem of Galois theory. We then give some applications of the theorem and in the next section our fourth proof of the Fundamental Theorem of Algebra.

Theorem 7.5.1
(Fundamental Theorem of Galois Theory) Let K be a Galois extension of F with Galois group $G = \text{Gal}(K/F)$. For each intermediate field E let $\tau(E)$ be the subgroup of G fixing E. Then:

(1) *τ is a bijection between intermediate fields containing F and subgroups of G.*
(2) *If H is a subgroup of G and $E = K^H$, then $\tau(E) = H$.*
(3) *K is Galois over E, and $\text{Gal}(K/E) = \tau(E)$.*
(4) *$|G| = |K:F|$.*
(5) *$|E:F| = |G:\tau(E)|$. That is, the degree of an intermediate field over the ground field is the index of the corresponding subgroup in the Galois group.*
(6) *E is Galois over F if and only if $\tau(E) \triangleleft G$. In this case*

$$\text{Gal}(E/F) \cong G/\tau(E) \cong \text{Gal}(K/F)/\text{Gal}(K/E).$$

(7) *The lattice of subfields of K containing F is the inverted lattice of subgroups of $\text{Gal}(K/F)$.*

Parts (1) through (6) have already been discussed. We say a little more about part (7). By the lattice of subfields of K over F we mean the complete collection of intermediate fields and their interrelationships. Similarly for the lattice of subgroups. One is the inverted lattice of the other, since in the field case if E is a subfield, the degree $|K : E|$ is the order of $\text{Gal}(K/E)$, while its index is the degree of $|E : F|$. Therefore, the subfields are from above in the lattice while the subgroups are from below. That is, we have the following diagram, where $\text{Gal}(K/K) = \{1\}$, the trivial group.

$$K \longleftrightarrow \text{Gal}(K/K)$$
$$\cup \quad\quad \cap$$
$$E \longleftrightarrow \text{Gal}(K/E)$$
$$\cup \quad\quad \cap$$
$$F \longleftrightarrow \text{Gal}(K/F)$$

We then illustrate this with three examples.

EXAMPLE 7.5.1
Consider the splitting field K of $x^4 + 1$ over $\mathbb{Q}$. As we saw in example 7.4.1, $K = \mathbb{Q}(i, \sqrt{2})$. Then $|K : \mathbb{Q}| = 4$, and the Galois group is $\mathbb{Z}_2 \times \mathbb{Z}_2$. There are then four automorphisms in $\text{Gal}(K/\mathbb{Q})$, given by:

$$1 : i \to i, \sqrt{2} \to \sqrt{2},$$
$$\sigma : i \to i, \sqrt{2} \to -\sqrt{2},$$
$$\tau : i \to -i, \sqrt{2} \to \sqrt{2},$$
$$\sigma\tau : i \to -i, \sqrt{2} \to -\sqrt{2}.$$

Each of these has order 2, and therefore there are five total subgroups of $G = \text{Gal}(K/\mathbb{Q})$, namely,

$$\{1\}, H_1 = \{1, \tau\}, H_2 = \{1, \sigma\tau\}, H_3 = \{1, \sigma\}, G.$$

We exhibit the five intermediate fields. The fixed field of G, is precisely $\mathbb{Q}$ while the fixed field of $\{1\}$ is all of K. Now, consider H_3. Since $i \in K$ and $\mathbb{Q}(i) \subset K$, $|K : \mathbb{Q}(i)| = 2$, so $\mathbb{Q}(i)$ will correspond to a subgroup of index 2 and thus order 2. Now, σ fixes i but not $\sqrt{2}$. Therefore, σ fixes $\mathbb{Q}(i)$ but not all of K, so the fixed field of H_3 is $\mathbb{Q}(i)$.

Similarly, $\mathbb{Q}(\sqrt{2}) \subset K$, and then $\mathbb{Q}(\sqrt{2})$ is the fixed field of H_1.

Finally, since $i, \sqrt{2} \in K$, it follows that $i\sqrt{2} \in K$, so $\mathbb{Q}(i\sqrt{2}) \subset K$. Since $\text{irr}(i\sqrt{2}, \mathbb{Q}) = x^2 + 2$, $|\mathbb{Q}(i\sqrt{2}) : \mathbb{Q}| = 2$. The automorphism $\sigma\tau$ fixes $i\sqrt{2}$; hence $\mathbb{Q}(i\sqrt{2})$ is the fixed field of H_2. We illustrate these relationships in Figures 7.1 and 7.2. □

7.5. The Fundamental Theorem of Galois Theory

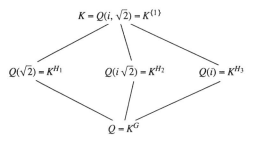

Lattice of Subfields

Figure 7.1.

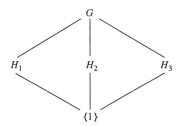

Lattice of Subgroups

Figure 7.2.

EXAMPLE 7.5.2
Let $K = \mathbb{Q}(\sqrt{2}, \sqrt{3})$. This is the splitting field of $(x^2 - 2)(x^2 - 3)$ over $\mathbb{Q}$, so K is Galois over $\mathbb{Q}$. Since $|\mathbb{Q}(\sqrt{2}) : \mathbb{Q}| = 2$ and $\sqrt{3} \notin \mathbb{Q}(\sqrt{2})$, we have $|K : \mathbb{Q}| = 4$. The Galois group can then be described by the four automorphisms

$$1 : \sqrt{2} \to \sqrt{2}, \sqrt{3} \to \sqrt{3},$$
$$\sigma : \sqrt{2} \to \sqrt{2}, \sqrt{3} \to -\sqrt{3},$$
$$\tau : \sqrt{2} \to -\sqrt{2}, \sqrt{3} \to \sqrt{3},$$
$$\sigma\tau : \sqrt{2} \to -\sqrt{2}, \sqrt{3} \to -\sqrt{3}.$$

It is easy to see that this group is $\mathbb{Z}_2 \times \mathbb{Z}_2$ and is isomorphic to the Galois group of the previous example. Thus different field extensions of the same ground field can have isomorphic Galois groups. The corresponding subgroups and fixed fields are given below.

$H_0 = \{1\} \Rightarrow$ fixed field is $K = \mathbb{Q}(\sqrt{2}, \sqrt{3})$;

$H_1 = \{1, \sigma\} \Rightarrow$ fixed field is $\mathbb{Q}(\sqrt{2})$;

$H_2 = \{1, \tau\} \Rightarrow$ fixed field is $\mathbb{Q}(\sqrt{3})$;

$H_3 = \{1, \sigma\tau\} \Rightarrow$ fixed field is $\mathbb{Q}(\sqrt{2}\sqrt{3}) = \mathbb{Q}(\sqrt{6})$;

$H_4 = G \Rightarrow$ fixed field is $\mathbb{Q}$

$\square$

EXAMPLE 7.5.3
As a somewhat more complicated example, consider the splitting field K of $x^4 - 2$ over $\mathbb{Q}$.

Over $\mathbb{C}$, $x^4 - 2 = (x^2 - \sqrt{2})(x^2 + \sqrt{2})$, so if $\omega = 2^{1/4}$, the four roots are $\omega, i\omega, -\omega, -i\omega$. Therefore, $i\omega/\omega = i \in K$ so $\mathbb{Q}(i, \omega) \subset K$. But $x^4 - 2$ splits in $\mathbb{Q}(i, \omega)$, so $K = \mathbb{Q}(i, \omega)$.

Now, $\mathbb{Q}(\omega)$ has degree 4 over $\mathbb{Q}$, and $i \notin \mathbb{Q}(\omega)$, since $\omega \in \mathbb{R}$. Therefore, $|K : \mathbb{Q}(\omega)| = 2$, since we are adjoining i, and hence $|K : \mathbb{Q}| = 8$.

The four conjugates of ω are $\omega, i\omega, -\omega, -i\omega$, while the conjugates of i are $\pm i$. Therefore, the eight automorphisms in $\text{Gal}(K/\mathbb{Q})$ are

$$1 : \omega \to \omega, i \to i,$$

$$\sigma : \omega \to i\omega, i \to i,$$

$$\sigma^2 : \omega \to -\omega, i \to i,$$

$$\sigma^3 : \omega \to -i\omega, i \to i,$$

$$\tau : \omega \to \omega, i \to -i,$$

$$\sigma\tau : \omega \to i\omega, i \to -i,$$

$$\sigma^2\tau : \omega \to -\omega, i \to -i,$$

$$\sigma^3\tau : \omega \to -i\omega, i \to -i.$$

A computation shows that $\sigma^4 = 1$, $\tau^2 = 1$, and $\sigma\tau = \tau\sigma^3$. The group is then D_4, the dihedral group of order 8 that represents the symmetries of a square. This group has the presentation

$$D_4 = <\sigma, \tau; \sigma^4 = 1, \tau^2 = 1, \sigma\tau = \tau\sigma^3>.$$

There are ten total subgroups of D_4. We list them below with the corresponding fixed fields. Notice that $\omega^2 = \sqrt{2} \in K$, so $\mathbb{Q}(i, \sqrt{2}) \subset K$.

$H_1 = \{1\} \Rightarrow$ fixed field is $K = \mathbb{Q}(i, \omega)$

$H_2 = \{1, \sigma, \sigma^2, \sigma^3\} \Rightarrow$ fixed field is $\mathbb{Q}(i)$,

$H_3 = \{1, \sigma^2\} \Rightarrow$ fixed field is $\mathbb{Q}(i, \sqrt{2})$,

$H_4 = \{1, \tau\} \Rightarrow$ fixed field is $\mathbb{Q}(\omega)$,

$H_5 = \{1, \sigma\tau\} \Rightarrow$ fixed field is $\mathbb{Q}(\omega + i\omega)$,

$H_6 = \{1, \sigma^2\tau\} \Rightarrow$ fixed field is $\mathbb{Q}(i\omega)$,

$H_7 = \{1, \sigma^3\tau\} \Rightarrow$ fixed field is $\mathbb{Q}(\omega - i\omega)$,

$H_8 = \{1, \sigma^2, \tau, \sigma^2\tau\} \Rightarrow$ fixed field is $\mathbb{Q}(\sqrt{2})$,

$H_9 = \{1, \sigma^2, \sigma\tau, \sigma^3\tau\} \Rightarrow$ fixed field is $\mathbb{Q}(i\sqrt{2})$,

$H_{10} = D_4 \Rightarrow$ fixed field is $\mathbb{Q}$. □

7.6 The Fundamental Theorem of Algebra – Proof Four

We now come to our fourth proof of the Fundamental Theorem of Algebra. This will use the Galois theory machinery built up in the previous sections.

Theorem 7.6.1
(Fundamental Theorem of Algebra) The complex number field $\mathbb{C}$ is algebraically closed. That is, any nonconstant complex polynomial has a root in $\mathbb{C}$.

Proof
If $f(x) \in \mathbb{C}[x]$, we can form the splitting field K for $f(x)$ over $\mathbb{C}$. This will be a Galois extension of $\mathbb{C}$ and thus a Galois extension of $\mathbb{R}$, since $\mathbb{C}$ is a finite extension of $\mathbb{R}$. The Fundamental Theorem of Algebra asserts that K must be $\mathbb{C}$ itself, and hence the Fundamental Theorem of Algebra is equivalent to the fact that any nontrivial Galois extension of $\mathbb{C}$ must be $\mathbb{C}$.

Let K be any finite extension of $\mathbb{R}$ with $|K : \mathbb{R}| = 2^m q, (2, q) = 1$. If $m = 0$, then K is an odd-degree extension of $\mathbb{R}$. Since K is separable over $\mathbb{R}$, it is a simple extension, and hence $K = \mathbb{R}(\alpha)$, where $irr(\alpha, \mathbb{R})$ has odd degree. However, odd-degree real polynomials always have a real root, and therefore $irr(\alpha, \mathbb{R})$ is irreducible only if its degree is one. But then $\alpha \in \mathbb{R}$ and $K = \mathbb{R}$. Therefore, if K is a nontrivial extension of $\mathbb{R}, m > 0$. This shows more generally that there are no odd-degree finite extensions of $\mathbb{R}$.

Suppose that K is a degree 2 extension of $\mathbb{C}$. Then $K = \mathbb{C}(\alpha)$ with deg $irr(\alpha, \mathbb{C}) = 2$. But complex quadratic polynomials always have roots in $\mathbb{C}$ so a contradiction. Therefore, $\mathbb{C}$ has no degree 2 extensions.

Now, let K be a Galois extension of $\mathbb{C}$. Then K is also Galois over $\mathbb{R}$. Suppose $|K : \mathbb{R}| = 2^m q, (2, q) = 1$. From the argument above we must have $m > 0$. Let $G = \text{Gal}(K/\mathbb{R})$ be the Galois group. Then $|G| = 2^m q, m > 0, (2, q) = 1$. Thus G has a 2-Sylow subgroup of order 2^m and index q. This would correspond to an intermediate field E with $|K : E| = 2^m$ and $|E : \mathbb{R}| = q$. However, then E is an odd-degree finite extension of $\mathbb{R}$. It follows that $q = 1$ and $E = \mathbb{R}$. Therefore, $|K : \mathbb{R}| = 2^m$ and $|G| = 2^m$.

Now, $|K : \mathbb{C}| = 2^{m-1}$ and suppose $G_1 = \text{Gal}(K/\mathbb{C})$. This is a 2-group. If it were not trivial, then from Lemma 7.2.4 there would exist a subgroup of order 2^{m-2} and index 2. This would correspond to an intermediate field E of degree 2 over $\mathbb{C}$. However from the argument above $\mathbb{C}$ has no degree 2 extensions. It follows then that G_1 is trivial, that is, $|G_1| = 1$, so $|K : \mathbb{C}| = 1$ and $K = \mathbb{C}$ completing the proof. ∎

As in our previous proofs, we have actually proved a more general result. In the above proof, outside of Galois theory, we used two facts: odd-degree real polynomials always have real roots, and degree 2 complex polynomials have complex roots. Using these two facts as properties we could prove the following generalization.

Theorem 7.6.2
Let K be an ordered field in which all positive elements have squareroots. Suppose further that each odd-degree polynomial in $K[x]$ has a root in K. Then $K(i)$ is algebraically closed, where $i = \sqrt{-1}$ is a root of the irreducible polynomial $x^2 + 1 \in K[x]$.

7.7 Some Additional Applications of Galois Theory

Galois theory was developed primarily as a tool for handling the proof of the insolvability by radicals of quintic polynomials. In this section we outline this proof. As mentioned in Chapter 1, the problem of solvability by radicals has a long and interesting history. The ability to solve quadratic equations and in essence the quadratic formula was known to the Babylonians some 3600 years ago. With the discovery of imaginary numbers the quadratic formula then says that any degree-two polynomial over $\mathbb{C}$ has a root in $\mathbb{C}$. In the sixteenth century the Italian mathematician Niccolo Tartaglia discovered a similar formula in terms of radicals to solve cubic equations. This **cubic formula** is now known erroneously as **Cardano's formula** in honor of Cardano, who first published it in 1545. An earlier special version of this formula was discovered by Scipione del Ferro. Cardano's student Ferrari extended the formula to solutions by radicals for fourth-degree polynomials.

From Cardano's work until the very early nineteenth century, attempts were made to find similar formulas for degree 5 polynomials. In 1805 Ruffini proved that fifth degree polynomial equations are insolvable by radicals in general – thus there exists no comparable formula for degree 5. Abel in 1825–1826 and Galois in 1831 extended Ruffini's result and proved the insolubility by radicals for all degrees 5 or greater.

7.7. Some Additional Applications of Galois Theory

In order to apply the Galois theory we must translate solvability by radicals into a group property, that is, a property that must be satisfied by the corresponding Galois group.

Definition 7.7.1
G is a **solvable group** if it has a finite series of subgroups

$$G = G_1 \supset G_2 \supset G_3 \ldots \supset G_n = 1,$$

with $G_{i+1} \triangleleft G_i$ and G_i/G_{i+1} abelian. Such a series is called a **normal series** for G, and the terms G_i/G_{i+1} are called the **factors** of the series. The definition can then be put concisely as: A group G is solvable if it has a normal series with abelian factors.

It can be shown that the class of solvable groups is closed under subgroups, factor groups, and finite direct products. That is if G, H, are solvable groups, then so are any subgroups of G or H; any factor groups of G or H, and the direct product $G \times H$.

Now, we must determine what is meant by solvability by radicals in terms of field extensions.

Definition 7.7.2
K is an **extension of** F **by radicals** if there exist elements $\alpha_1, \ldots, \alpha_r \in K$ and integers $n_1, \ldots, n_r$ such that $K = F(\alpha_1, \ldots, \alpha_r)$, with $\alpha_1^{n_1} \in F$, and $\alpha_i^{n_i} \in F(\alpha_1, \ldots, \alpha_{i-1})$, for $i = 2, \ldots, r$. A polynomial $f(x) \in F[x]$ is **solvable by radicals over** F if the splitting field K of $f(x)$ over F is an extension by radicals.

The key result tying these two concepts – solvability of groups and solvability by radicals – is the following.

Theorem 7.7.1
Suppose K is a Galois extension of F with char F = 0. Then if K is an extension of F by radicals, Gal(K/F) is a solvable group.

Therefore, to show that it is not possible in general to solve a polynomial of degree 5 or greater by radicals, we must show that for any $n \geq 5$ there exists a polynomial of degree n whose Galois group is not solvable. The Galois group is always contained in a symmetric group, and the following can be proved.

Theorem 7.7.2
For any $n \geq 5$ the symmetric group S_n is not a solvable group.

Therefore, we could show the insolvability by radicals by exhibiting for each n a polynomial whose Galois group is the whole symmetric group S_n. This is what is done.

Let $y_1, \ldots, y_n$ be independent transcendental elements over $\mathbb{Q}$ and let $K = \mathbb{Q}(y_1, \ldots, y_n)$. Let $s_1, \ldots, s_n$ be the elementary symmetric polynomials in $y_1, \ldots, y_n$ (see Section 6.4) and let $F = \mathbb{Q}(s_1, \ldots, s_n)$. Then K is a Galois extension of F and $\text{Gal}(K/F) \cong S_n$.

Theorem 7.7.3
Let $y_1, \ldots, y_n$ be independent transcendental elements over $\mathbb{Q}$. Then the polynomial $f(x) = \prod_{i=1}^{n}(x - y_i)$ is not solvable by radicals over $F = \mathbb{Q}(s_1, \ldots, s_n)$, where $s_1, \ldots, s_n$ are the elementary symmetric polynomials in $y_1, \ldots, y_n$.

As a final application we indicate the impossibility of certain geometric ruler (straightedge) and compass constructions. Greek mathematicians in the classical period posed the problem of finding geometric constructions using only ruler and compass to double a cube, trisect an angle, and square a circle. The Greeks were never able to prove that such constructions were impossible but were able to construct solutions to these problems only using other techniques, including conic sections. In 1837, Pierre Wantzel, using algebraic methods was able to prove the impossibility of either trisecting an angle or doubling a cube. With the proof that π is transcendental (done by Lindemann in 1882 – see Section 6.6) Wantzel's method could be applied to showing that squaring the circle is also impossible. We will outline the algebraic method. As in the insolvability of the quintic, we must translate into the language of field extensions. As a first step we define a **constructible number**.

Definition 7.7.3
Suppose we are given a line segment of unit length. An $\alpha \in \mathbb{R}$ is **constructible** if we can construct a line segment of length $|\alpha|$ in a finite number of steps from the unit segment using a ruler and compass.

Recall from elementary geometry that using a ruler and compass it is possible to draw a line parallel to a given line segment through a given point, to extend a given line segment, and to erect a perpendicular to a given line at a given point on that line. Our first result is that the set of all constructible numbers forms a subfield of $\mathbb{R}$.

Theorem 7.7.4
The set C of all constructible numbers forms a subfield of $\mathbb{R}$. Further, $\mathbb{Q} \subset C$.

Proof
Let C be the set of all constructible numbers. Since the given unit length segment is constructible, we have $1 \in C$. Therefore, $C \neq \emptyset$, and thus

7.7. Some Additional Applications of Galois Theory

to show that it is a field we must show that it is closed under the field operations.

Suppose α, β are constructible. We must show then that $\alpha \pm \beta$, $\alpha\beta$, and α/β for $\beta \neq 0$ are constructible. If $\alpha, \beta > 0$, construct a line segment of length $|\alpha|$. At one end of this line segment extend it by a segment of length $|\beta|$. This will construct a segment of length $\alpha + \beta$. Similarly, if $\alpha > \beta$, lay off a segment of length $|\beta|$ at the beginning of a segment of length $|\alpha|$. The remaining piece will be $\alpha - \beta$. By considering cases we can do this in the same manner if either α or β or both are negative. These constructions are pictured in Figure 7.3. Therefore, $\alpha \pm \beta$ are constructible.

Figure 7.3. Constructibility of $\alpha \pm \beta$

In Figure 7.4 we show how to construct $\alpha\beta$. Let the line segment $\overline{OA}$ have length $|\alpha|$. Consider a line L through O not coincident with $\overline{OA}$. Let $\overline{OB}$ have length $|\beta|$ as in the diagram. Let P be on ray $\overline{OB}$ so that $\overline{OP}$ has length 1. Draw $\overline{AP}$ and then find Q on ray $\overline{OA}$ such that $\overline{BQ}$ is parallel to $\overline{AP}$. From similar triangles we then have

$$\frac{|\overline{OP}|}{|\overline{OB}|} = \frac{|\overline{OA}|}{|\overline{OQ}|} \Rightarrow \frac{1}{|\beta|} = \frac{|\alpha|}{|\overline{OQ}|}.$$

Then $|\overline{OQ}| = |\alpha||\beta|$, and so $\alpha\beta$ is constructible.

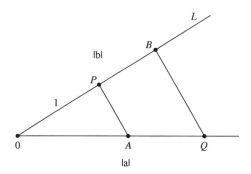

Figure 7.4. Constructibility of $\alpha\beta$

A similar construction, pictured in Figure 7.5, shows that α/β for $\beta \neq 0$ is constructible. Find $\overline{OA}, \overline{OB}, \overline{OP}$ as above. Now, connect A to B and let $\overline{PQ}$ be parallel to $\overline{AB}$. From similar triangles again we have

$$\frac{1}{|\beta|} = \frac{|\overline{OQ}|}{|\alpha|} \Rightarrow \frac{|\alpha|}{|\beta|} = |\overline{OQ}|.$$

Hence α/β is constructible.

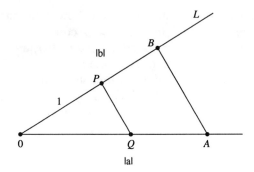

Figure 7.5. Constructibility of α/β

Therefore, C is a subfield of $\mathbb{R}$. Since char $C = 0$, it follows that $\mathbb{Q} \subset C$. ∎

Let us now consider analytically how a constructible number is found in the plane. Starting at the origin and using the unit length and the constructions above, we can locate any point in the plane with rational coordinates. That is, we can construct the point $P = (q_1, q_2)$ with $q_1, q_2 \in \mathbb{Q}$. Using only ruler and compass, any further point in the plane can be determined in one of the following three ways.

(1) The intersection point of two lines each of which passes through two known points each having rational coordinates.
(2) The intersection point of a line passing through two known points having rational coordinates and a circle whose center has rational coordinates and whose radius squared is rational.
(3) The intersection point of two circles each of whose centers has rational coordinates and each of whose radii is the square root of a rational number.

Analytically, the first case involves the solution of a pair of linear equations each with rational coefficients and thus only leads to other rational numbers. In cases two and three we must solve equations of the form $x^2 + y^2 + ax + by + c = 0$, with $a, b, c \in \mathbb{Q}$. These will then be quadratic equations over $\mathbb{Q}$, and thus the solutions will either be in $\mathbb{Q}$ or in a quadratic extension $\mathbb{Q}(\sqrt{\alpha})$ of $\mathbb{Q}$. Once a real quadratic extension of $\mathbb{Q}$ is found, the process can be iterated. Conversely it can be shown that if α is constructible, so is $\sqrt{\alpha}$. We thus can prove the following theorem.

Theorem 7.7.5
If γ is constructible with $\gamma \notin \mathbb{Q}$, then there exists a finite number of elements $\alpha_1, \ldots, \alpha_r \in \mathbb{R}$ with $\alpha_r = \gamma$ such that for $i = 1, \ldots, r, \mathbb{Q}(\alpha_1, \ldots, \alpha_i)$ is a

7.7. Some Additional Applications of Galois Theory

quadratic extension of $\mathbb{Q}(\alpha_1, \ldots, \alpha_{i-1})$. *In particular,* $|\mathbb{Q}(\gamma):\mathbb{Q}| = 2^n$ *for some* $n \geq 1$.

Therefore, the constructible numbers are precisely those real numbers that are contained in repeated quadratic extensions of $\mathbb{Q}$. We now use this idea to show the impossibility of the three mentioned construction problems.

EXAMPLE 7.7.1
It is impossible to **double the cube**. This means that it is impossible, given a cube of given side length, to construct using a ruler and compass, a side of a cube having double the volume of the original cube.

Let the given side length be 1, so that the original volume is also 1. To double this we would have to construct a side of length $2^{1/3}$. However $|\mathbb{Q}(2^{1/3}) : \mathbb{Q}| = 3$ since $irr(2^{1/3}, \mathbb{Q}) = x^3 - 2$. This is not a power of 2 so $2^{1/3}$ is not constructible. □

EXAMPLE 7.7.2
It is impossible to **trisect an angle**. This means that it is impossible in general to trisect a given angle using only a ruler and compass.

An angle θ is constructible if and only if a segment of length $|\cos\theta|$ is constructible. Since $\cos(\pi/3) = 1/2$, $\pi/3$ is constructible. We show that it cannot be trisected by ruler and compass.

The following trigonometric identity can be proved

$$\cos(3\theta) = 4\cos^3(\theta) - 3\cos(\theta).$$

Let $\alpha = \cos(\pi/9)$. From the above identity we have $4\alpha^3 - 3\alpha - \frac{1}{2} = 0$. The polynomial $4x^3 - 3x - \frac{1}{2}$ is irreducible over $\mathbb{Q}$ so $irr(\alpha, \mathbb{Q}) = x^3 - \frac{3}{4}x - \frac{1}{8}$. It follows that $|\mathbb{Q}(\alpha) : \mathbb{Q}| = 3$, and hence α is not constructible. Therefore, the corresponding angle $\pi/9$ is not constructible. Therefore, $\pi/3$ is constructible, but it cannot be trisected. □

EXAMPLE 7.7.3
It is impossible to **square the circle**. That is it is impossible in general, given a circle, to construct using ruler and compass a square having area equal to that of the given circle.

Suppose the given circle has radius 1. It is then constructible and would have an area of π. A corresponding square would then have to have a side of length $\sqrt{\pi}$. But π is transcendental, so $\sqrt{\pi}$ is not constructible. □

The other great construction problem solved by Galois theory was the construction of regular n-gons. A regular n-gon will be constructible for $n \geq 3$ if and only if the angle $2\pi/n$ is constructible, which is the case if and only if $\cos 2\pi/n$ is constructible. The algebraic study of the constructibility of regular n-gons was initiated by Gauss in the early part of the nineteenth century.

7.8 Algebraic Extensions of $\mathbb{R}$ and Concluding Remarks

Suppose F is a finite field extension of the real numbers $\mathbb{R}$. From the proof of the Fundamental Theorem of Algebra it follows that either $F = \mathbb{R}$ or $|F : \mathbb{R}| = 2$ and $F \cong \mathbb{C}$. Thus the Fundamental Theorem of Algebra can also be phrased in the following way.

Theorem 7.8.1
If F is a field extension of $\mathbb{R}$ with $|F:\mathbb{R}| > 1$, then $|F:\mathbb{R}| = 2$ and $F \cong \mathbb{C}$.

If we extend somewhat the definition of a field we can obtain a further classification of extensions of the reals. A **division ring** is an algebraic structure with the same properties as a field but in which multiplication is not necessarily commutative. That is, it is a ring with an identity where every nonzero element has a multiplicative inverse. Thus every field is a division ring but as we will see below there are noncommutative division rings. A noncommutative division ring is called a **skew field**. If F is a field and $F \subset D$ where D is a division ring, then D is still a vector space over F. Further there is a multiplication in D so that each nonzero element has a multiplicative inverse. The identity in F must be an identity for the division ring. In this case D is called a **division algebra** over F. Here the elements of F will commute with all elements in D.

We now give a method to construct a class of skew fields. Suppose F is a field in which no sum of squares can be zero. That is, if $x_1^2 + \cdots + x_n^2 = 0$ with $x_1, \ldots, x_n \in F$ then $x_i = 0$ for $i = 1, \ldots, n$. A field satisfying this is called a **totally real field**. Consider now a vector space H_F of dimension 4 over F with basis $1, i, j, k$. We identify 1 with the identity in F and then build a multiplication on H_F by defining the products of the basis elements. Let $i^2 = j^2 = k^2 = -1$ and $ijk = -1$. This will completely define, using associativity, all products of basis elements. For example, from $ijk = -1$ we have $ij = -k^{-1} = k$ since $k^2 = -1$. Then $ik = i(ij) = (ii)j = -j$. It is easy to show that $ij = -ji$, so that this product is noncommutative.

A general element of H_F has the form $f_0 + f_1 i + f_2 j + f_3 k$. Multiplication of elements like this is done by algebraic manipulation using the defined products of basis elements. We thus get a product on H_F. It is a straightforward computation to prove the following theorem.

7.8. Algebraic Extensions of $\mathbb{R}$ and Concluding Remarks

Theorem 7.8.2
*For a totally real field F, H_F forms a division algebra of degree 4 over F. H_F is called the **quaternion algebra** over F.*

The only difficulty in the proof is to show the existence of inverses. This is done just as in the complex numbers and we leave it to the exercises.

The **quaternions** H are the quaternion algebra over the reals. That is, $H = H_{\mathbb{R}}$. This algebra consists of all elements $r_0 + r_1 i + r_2 j + r_3 k$ with $r_i \in \mathbb{R}$. Identifying $\mathbb{R}$ with the first component we get that $\mathbb{R} \subset H$. Therefore, H is a finite *skew* field extension of $\mathbb{R}$. The following theorem says that this is the only one.

Theorem 7.8.3
Let D be a finite-dimensional division algebra over $\mathbb{R}$ of degree greater than 1. If D is a field, then $|D:\mathbb{R}| = 2$ and $D \cong \mathbb{C}$. If D is a skew field, then $|D:\mathbb{R}| = 4$ and $D \cong H$.

Proof
If D is a field the result is just the reformulation of the Fundamental Theorem of Algebra so we consider the case where D is a skew field. We outline the proof below and leave the details to the exercises.

(1) If D is a skew field extension of $\mathbb{R}$, it must have degree at least 4. As in the field case every element is a root of a polynomial with coefficients in $\mathbb{R}$. If the degree is 3, there is a root in $\mathbb{R}$, and if the degree is 2, the resulting extension must be commutative and thus isomorphic to $\mathbb{C}$.

(2) If D has degree four over $\mathbb{R}$, then it has a basis $1, e_1, e_2, e_3$. We can show that the nonreal basis elements must satisfy quadratic polynomials over $\mathbb{R}$ and thus can be chosen to behave like the imaginary unit i. Thus $e_1^2 = e_2^2 = e_3^2 = -1$. By the noncommutativity of D we can obtain that $e_1 e_2 = e_3$ and thus $D \cong H$.

(3) If D is a division algebra over $\mathbb{R}$ of degree $n + 1$, then as above there is a basis $1, e_1, \ldots, e_n$ with $e_i^2 = -1, i = 1, \ldots, n$. We can then show that for each pair i, j with $i \neq j$ we have $e_i e_j + e_j e_i \in \mathbb{R}$. From this we can then show that we can replace e_2 by e_2' with $e_1 e_2' + e_2' e_1 = 0$. (If $e_1 e_2 + e_2 e_1 = 2c \in \mathbb{R}$ set $e_2' = (e_2 + ce_1)/\sqrt{1 - c^2}$.) We can then show that this leads to a contradiction if $n > 3$. (We may choose $e_3 = e_1 e_2$. Then let $a_{ij} = e_i e_j + e_j e_i$, and so $a_{12} = 0$. Then $-2e_4 = a_{14} e_1 + a_{24} e_2 + a_{34} e_3$, violating the independence of the basis). ∎

We have now completed our algebraic look at the Fundamental Theorem of Algebra. In this approach the theorem was really saying something about algebraic extensions of the reals. Although the proofs were algebraic, they were still dependent on the continuity properties of real polynomials and hence on the completeness of the reals. In the next two chapters we examine two topologically motivated proofs of the Fundamental Theorem.

Exercises

7.1. Prove Theorem 7.2.1. If R is any algebraic structure, then $\mathrm{Aut}(R)$ forms a group. Further, if S is a substructure, then $\mathrm{Aut}(S)$ is a subgroup of $\mathrm{Aut}(R)$.

7.2. Let G be a cyclic group of order 6. Determine the structure of $\mathrm{Aut}(G)$.

7.3. If G is a finite group and H is a subgroup, show that $N_G(H)$ is a subgroup and $|G : N_G(H)|$ is the number of distinct conjugates of H in G.

7.4. If G is abelian and H is a subgroup, show that G/H is also abelian.

7.5. Suppose p, q are distinct primes with $p > q$. If G is a group of order pq, then G has a normal p-Sylow subgroup.

7.6. Let F be a finite field. Show that F has p^n elements for some prime p. (Hint: Since $|F| < \infty$ char $F \neq 0$ so its prime subfield is $\mathbb{Z}_p$. F is then a vector space over $\mathbb{Z}_p$).

7.7. (a) Let $\alpha = i + \sqrt{2}$. What is $irr(\alpha, \mathbb{Q})$ and $|\mathbb{Q}(\alpha) : \mathbb{Q}|$?
     (b) Let $\alpha = \sqrt{3} + \sqrt{7}$. What is $irr(\alpha, \mathbb{Q})$ and $|\mathbb{Q}(\alpha) : \mathbb{Q}|$?

7.8. Show that $\mathbb{Q}(\sqrt{3} + \sqrt{7}) = \mathbb{Q}(\sqrt{3}, \sqrt{7})$. (Hint: Compute the degrees).

7.9. Suppose $|E : F| = n$ and $p(x) \in F[x]$ is irreducible over F. Then if $\deg p(x)$ and n are relatively prime then $p(x)$ has no roots in E.

7.10. Suppose $|E : \mathbb{R}| < \infty$, then $E = \mathbb{R}$ or $E = \mathbb{C}$.

7.11. Determine the Galois group and the lattice of subfields for the splitting field of $x^3 - 6$ over $\mathbb{Q}$.

7.12. Let G be a finite group. Show that $G = \mathrm{Gal}(K/F)$ for some field F and Galois extension K.

7.13. If G is a group and $g_1, g_2 \in G$, then the **commutator** of g_1, g_2, denoted by $[g_1, g_2]$ is $g_1 g_2 g_1^{-1} g_2^{-1}$. The commutator subgroup of G, denoted by G' is the subgroup generated by all commutators.
     (a) Prove that G' is a normal subgroup of G.
     (b) Prove that G/G' is abelian.
     (c) Show that if G/H is abelian then $G' \subset H$.

7.14. If G is solvable and H is a subgroup of G, then H is also solvable. If H is normal in G, then G/H is solvable.

7.15. Let $\alpha = r_0 + r_1 i + r_2 j + r_3 k$ be a real quaternion, that is, $\alpha \in H$. Define its conjugate $\bar{\alpha} = r_0 - r_1 i + r_2 j + r_3 k$. Show that $\alpha \bar{\alpha}$ is real and then use this to define an inverse for each nonzero quaternion.

7.16. Fill in the details that for a totally real field F, H_F forms a division algebra.

Exercises

7.17. Fill in the details of the proof of Theorem 7.8.3.

7.18. Let S_n be the symmetric group on n symbols. S_n then can be considered as the group of permutations on $\{1, 2, \ldots, n\}$. A permutation $\sigma \in S_n$ is a **cycle** of order k if there exists a subset $\{i_1, i_2, \ldots, i_k\}$ of $\{1, 2, \ldots, n\}$ such that $\sigma(i_1) = i_2, \sigma(i_2) = i_3, \ldots, \sigma(i_k) = i_1$ (σ starts at i_1 and then cycles through $\{i_1, i_2, \ldots, i_k\}$ and then back to i_1). For example the permutation $\begin{pmatrix} 3 & 4 & 6 \\ 4 & 6 & 3 \end{pmatrix}$ is a cycle of order 3 within S_6. We denote a cycle by starting with any integer in it and then listing the images in order. Thus the 3-cycle $\begin{pmatrix} 3 & 4 & 6 \\ 4 & 6 & 3 \end{pmatrix}$ would be expressed as (346)

(a) Prove that every permutation in S_n is a product of disjoint cycles. (Hint: Suppose $\sigma \in S_n$. Start with 1 and trace out what happens to $\sigma(1), \sigma(\sigma(1))$. Eventually it will have to cycle back to 1. Then go to the smallest integer left out of this cycle.) For example $\begin{pmatrix} 1 & 2 & 3 & 4 & 5 \\ 2 & 3 & 1 & 5 & 4 \end{pmatrix} = (123)(45)$

(b) A **transposition** is a cycle of order 2. Prove that every permutation in S_n can be expressed as a product of transpositions (not necessarily disjoint). (Hint: Show that every cycle can be written as a product of tranpositions.)

(c) Show that the set of permutations in S_n that can be written as a product of an even number of transpositions forms a subgroup of S_n of index 2. This subgroup is called the **alternating group** on n symbols denoted by A_n. In general a permutation is an **even** permutation if it can be expressed as an even number of transpositions and **odd** otherwise. Thus the alternating group is the subgroup of even permutations.

CHAPTER 8
Topology and Topological Spaces

8.1 Winding Number and Proof Five

We have now seen four different proofs of the Fundamental Theorem of Algebra. The first two were purely analysis, while the second pair involved a wide collection of algebraic ideas. However, we should realize that even in these algebraic proofs we did not totally leave analysis. Each of these proofs used the fact that odd-degree real polynomials have real roots. This fact is a consequence of the intermediate value theorem, which depends on continuity. Continuity is a topological property and we now proceed to our final pair of proofs, which involve topology.

Consider the curve

$$\gamma(t) = z_0 + re^{it}, 0 \le t \le 2n\pi.$$

Geometrically, this is a circle that winds n times around the point z_0. We make this more precise. If we integrate, then we get

$$\frac{1}{2\pi i}\int_\gamma \frac{dz}{z-z_0} = \frac{1}{2\pi i}\int_0^{2n\pi} \frac{ire^{it}}{re^{it}}dt = n.$$

The number n is called the **winding number** of the curve γ aound z_0. More generally, if γ is any closed continuously differentiable curve in $\mathbb{C}$ and $z_0 \in \mathbb{C} -$ image γ, then it can be shown that

$$n(\gamma, z_0) = \frac{1}{2\pi i}\int_\gamma \frac{dz}{z-z_0}$$

is an integer, called the **winding number** of γ around z_0.

8.1. Winding Number and Proof Five

Now we consider the function $g(z) = z^n$. Let C_r be the circle $z = re^{it}$, $0 \le t \le 2\pi$, of radius r about the origin. On this circle then, $z^n = r^n e^{int}$. As t runs from 0 to 2π, z winds once around the circle. At the same time, nt runs from 0 to $2n\pi$, and so z^n winds n times around the circle of radius r^n. We say that the function z^n has **winding number** n about the origin.

We now use this idea of winding number to fashion our fifth proof of the Fundamental Theorem of Algebra. First we must generalize the idea of the winding number of a function. If γ is a closed continuously differentiable curve and $f : \mathbb{C} \to \mathbb{C}$ is a continuous function, then $f(\gamma)$ is also a closed curve. We say f winds the closed curve γ, n times around z_0 if $f(\gamma)$ has winding number n around z_0. Most important for us is when γ is a circle.

Now, if $f(z)$ and $g(z)$ are sufficiently close on a circle C_r of radius r around the origin, then both $f(z)$ and $g(z)$ will wind C_r the same number of times about the origin. That is, if $|f(z) - g(z)| < \epsilon$ for some small enough ϵ on a circle of radius r, then $f(C_r)$, $g(C_r)$ have the same winding number around the origin.

Although the above fact can be made precise, we will just exhibit it by using what is now called the **fellow-traveler property**. Surprisingly, this idea has become very important in geometric group theory. Consider two travelers tied together by a rope. If one traveler $\{f(z)\}$ traverses a circle around the origin, the fellow-traveler $\{g(z)\}$ will also, following a different path, provided that the length of the rope $\{\epsilon\}$ is less than the radius of the circle. We illustrate this with a picture (Figure 8.1).

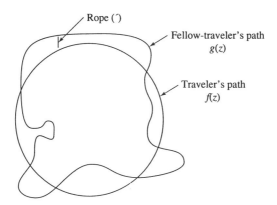

Figure 8.1. Fellow Traveler Property

We now give our fifth proof of the Fundamental Theorem of Algebra.

Theorem 8.1.1 (Fundamental Theorem of Algebra)
Any nonconstant complex polynomial has a complex root.

Proof

Suppose $f(z) = a_n z^n + \cdots + a_0$ with $a_n \neq 0$ and $n \geq 1$. In searching for a root we can without loss of generality assume that $a_n = 1$, so that

$$f(z) = z^n + a_{n-1} z^{n-1} + \cdots + a_0.$$

If $a_0 = 0$, then $z = 0$ is a root so we may assume that $a_0 \neq 0$.

Recall that $f(z)$ is then a continuous function $\mathbb{C} \to \mathbb{C}$. Further,

$$\lim_{z \to \infty} \frac{z^n}{f(z)} = 1,$$

and so for a sufficiently large circle C_r we have

$$|z^n - f(z)| \leq \lambda r^n \tag{8.1.1}$$

with $0 < \lambda < 1$ and z on C_r.

For any $r > 0$, z^n winds C_r around the origin n times. Therefore from the fellow-traveler property $f(z)$ will also wind a sufficiently large C_r also n times around the origin.

For a small enough radius r, $f(z) \approx a_0$ on C_r so $f(C_r)$ makes a small loop around a_0 and will not wind around the origin at all. Since $f(z)$ is continuous, $f(C_r)$ will depend continuously on r. Since $f(C_r)$ has winding number 0 for a small radius r and winds n times around the origin for larger r, there must be an intermediate radius r_1 with $f(C_{r_1})$ passing through the origin. It follows then that there must be a point z_0 on C_{r_1} with $f(z_0) = 0$, proving the theorem. ∎

In the next chapter we extend this proof to a general topological proof of the Fundamental Theorem of Algebra. To do this we must introduce some basic concepts and results from topology.

8.2 Topology – an Overview

Topology is the name of a major branch of mathematics that arises in large part from the so-called topological properties of the complex plane. Literally topology means the analysis or science of position and in older texts one finds the term **analysis situs** used synonymously with topology.

A topological property of the complex plane is one that involves continuity. These properties include such concepts as **open set, closed set, compact set,** and **connected set**. Topology deals then with properties preserved by continuous bijections, although of course a more general definition of continuity is necessary. Most generally, topology is then the study of **topological spaces**, which can be thought of as the most arbitrary spaces where continuity can be defined. We will make this precise in Section 8.4.

8.2. Topology – an Overview

Historically, topology has followed two principal developmental paths and approaches, which do overlap. The first of these paths is from geometry. In this approach a topological space is viewed as a generalized geometric configuration, with two such configurations being equivalent if they can be continuously deformed into each other. This led topology to obtain the title "rubber sheet geometry" with two objects being topologically the same if they can be manipulated into each other without tearing. Thus, a circle is in a sense topologically the same as any closed curve, and a torus (inner tube) is topologically the same as a sphere with a handle (jolly jumper). We picture these in Figure 8.2.

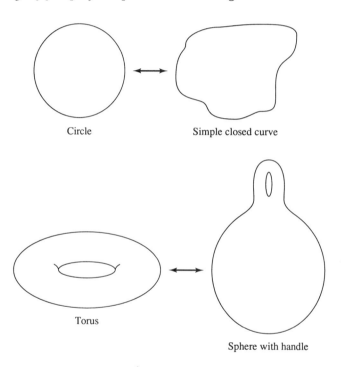

Figure 8.2.

Modern topology goes far beyond this rubber sheet geometric approach; tearing, cutting, and pasting are all permitted as long as they are done in a continuous manner. However, the ideas of rubber sheet geometry persist.

In this geometric approach the emphasis is on the spaces themselves. This approach has led to the development of homology and homotopy theory (discussed in the next chapter) and the topological theory of manifolds (spaces that look "locally" like Euclidean space).

The second path topology followed generalizes analysis. Here the primary emphasis is on continuous functions and their properties. The spaces themselves are regarded as just the domains of these functions. This approach has led to the development of "generalized" Euclidean spaces – Banach spaces and Hilbert spaces – over which a theory of differentiation and integration can be constructed.

Topology has many sub-branches and it breaks roughly into two broad subcategories. The first is called **point-set topology**. This deals with the study of topological spaces directly. It is this approach that we will follow in this chapter. The second is **algebraic topology**. Here topological spaces and structures are studied in terms of algebraic objects (groups and rings) associated with them. We will see this approach, which will lead to our final proof of the Fundamental Theorem of Algebra, in the next chapter.

8.3 Continuity and Metric Spaces

As indicated in the last section, topology and topological properties arise from the concept of continuity. Here we first review and then extend this concept.

Recall that a single variable real-valued function $y = f(x)$ is **continuous at a point** x_0 if for any $\epsilon > 0$ there exists a $\delta > 0$, (which depends on ϵ and x_0), such that $|f(x) - f(x_0)| < \epsilon$ whenever $0 \leq |x - x_0| < \delta$. The function $y = f(x)$ is continuous on an interval $[a, b]$ if it is continuous at each point of $[a, b]$. Notice that in the above definition, the absolute values are used to measure distance along the real line, so the definition can be rephrased in the following manner: Given $\epsilon > 0$ there exists a $\delta > 0$ such that the distance between $f(x)$ and $f(x_0)$ is less than ϵ whenever the distance between x and x_0 is less than δ.

As we saw for complex functions in Chapter 4, this definition can be extended almost verbatim to real-valued functions defined on $\mathbb{R}^2$. That is, let $P_0 = (x_0, y_0)$, then a real-valued function $z = f(x, y)$ is continuous at P_0 if for any $\epsilon > 0$ there exists a $\delta > 0$ (again depending on ϵ and P_0) such that $|f(x, y) - f(x_0, y_0)| < \epsilon$ whenever $|(x, y) - (x_0, y_0)| < \delta$.

The second absolute value above refers to distance in $\mathbb{R}^2$. Let us restate this in terms of distance, and this restatement will serve as a take-off point for extending the concept. First of all, in either the real line $\mathbb{R}$, or the plane $\mathbb{R}^2$, let $d(x, y)$ refer to the distance between the two points x, y. Then, a function $z = f(P)$ from $\mathbb{R}^2$ to $\mathbb{R}$ is continuous at $P_0 \in \mathbb{R}^2$ if for any $\epsilon > 0$ there exists a $\delta > 0$ such that $d(f(P), f(P_0)) < \epsilon$ whenever $d(P, P_0) < \delta$. (Note that the first d refers to distance in $\mathbb{R}$, while the second d is distance in $\mathbb{R}^2$.) Further, as for single-variable functions, $z = f(P)$ is continuous on some subset in $\mathbb{R}^2$ if it is continuous at each point of this subset.

8.3. Continuity and Metric Spaces

It is clear then that the definitions for continuity in $\mathbb{R}$ and $\mathbb{R}^2$ can be extended to any n-dimensional real space $\mathbb{R}^n$, provided that distance can be measured.

Definition 8.3.1
Let $P_1 = (x_1, \ldots, x_n), P_2 = (y_1, \ldots, y_n) \in \mathbb{R}^n$. Then the distance from P_1 to P_2, denoted by $d(P_1, P_2)$, is

$$d(P_1, P_2) = \sqrt{\sum_{i=1}^{n} (x_i - y_i)^2}.$$

Notice that this is just the generalization of the distance formula in $\mathbb{R}^2$. It is straightforward (see the exercises) to verify that this definition of distance will then satisfy all the basic properties ordinarily associated with distance.

Lemma 8.3.1
Let $d:\mathbb{R}^n \times \mathbb{R}^n \to \mathbb{R}$ be the distance function on $\mathbb{R}^n$ defined as above. Then:
(1) $d(P_1,P_2) \geq 0$ for all $P_1,P_2 \in \mathbb{R}^n$, and $d(P_1,P_2) = 0$ if and only if $P_1 = P_2$.
(2) $d(P_1,P_2) = d(P_2,P_1)$ for all $P_1,P_2 \in \mathbb{R}^n$.
(3) $d(P_1,P_2) \leq d(P_1,P_3) + d(P_3,P_2)$ for all $P_1,P_2,P_3 \in \mathbb{R}^n$ (*triangle inequality*).

Now that we have defined a distance function, we can define continuity. Further the range of a continuous function need not be in $\mathbb{R}$.

Definition 8.3.2
Let $f : \mathbb{R}^n \to \mathbb{R}^m$. Then f is **continuous at** $P_0 \in \mathbb{R}^n$ if for any $\epsilon > 0$ there exists a $\delta > 0$ such that $d(P, P_0) < \delta$ implies that $d(f(P), f(P_0)) < \epsilon$. The function f is **continuous on a subset** $U \subset \mathbb{R}^n$ if it is continuous at each point of U.

In all of the above discussion, the only necessity for defining continuity was the ability to measure distance. We now abstract this to consider general spaces where distance can be measured.

Definition 8.3.3
Let M be a nonempty set. A **metric**, or **distance function** on M is a function $d : M \times M \to \mathbb{R}$ satisfying

(1) $d(x, y) \geq 0$ for all $x, y \in M$, and $d(x, y) = 0$ if and only if $x = y$.
(2) $d(x, y) = d(y, x)$ for all pairs $x, y \in M$.
(3) $d(x, y) \leq d(x, z) + d(z, y)$ for all triples $x, y, z \in M$ (This is the **triangle inequality**).

A **metric space** is a pair (M, d) consisting of a set M and a metric d defined on it. The elements of M are called the **points** of the metric space.

EXAMPLE 8.3.1
(1) For any $n \geq 1$, $\mathbb{R}^n$, n-dimensional real space, forms a metric space with the metric defined as in Definition 8.3.1. $\mathbb{R}^n$ with this metric is called n-**dimensional Euclidean space.**

(2) Let $C^0[a, b]$ be the set of all continuous functions on the closed interval $[a, b]$. If $f(x), g(x) \in C^0[a, b]$, define

$$d(f(x), g(x)) = \left(\int_a^b |f(x) - g(x)|^2 dx \right)^{1/2}.$$

This then defines a metric on $C^0[a, b]$ (see the exercises), and so $C^0[a, b]$ equipped with this metric becomes a metric space.

(3) Let $C^0[a, b]$ be the set of all continuous functions on the closed interval $[a, b]$. Now, if $f(x), g(x) \in C^0[a, b]$, define

$$d(f(x), g(x)) = \max{}_{[a,b]} |f(x) - g(x)|.$$

This also defines a metric on $C^0[a, b]$, and so again $C^0[a, b]$ becomes a (different) metric space with this metric.

These last two examples show that a set may have more than one metric defined on it. In particular, any set M can be made into a metric space by defining $d(x, y) = 0$ if $x = y$ and $d(x, y) = 1$ if $x \neq y$. However, in most cases this is not a very interesting metric. □

The second two metric spaces in Example 8.3.1 are specific cases of a general type of space arising from certain types of real vector spaces. These play an important role in analysis so we briefly introduce them.

Definition 8.3.4
Let V be a vector space over the real numbers $\mathbb{R}$. Then an **inner product** on V is a function $<,>: V \times V \to \mathbb{R}$ satisfying

(1) $<x, x> \geq 0$ for all $x \in V$, and $<x, x> = 0$ if and only if $x = 0$ (the zero vector).
(2) $<x, y> = <y, x>$ for all pairs of vectors $x, y \in V$.
(3) $<x+y, z> = <x, z> + <y, z>$ for all triples of vectors $x, y, z \in V$.
(4) $<\alpha x, y> = \alpha <x, y>$ for $x, y \in V$ and $\alpha \in \mathbb{R}$.

An **inner product space** is a vector space V with an inner product defined on it.

A **norm** on V is a function $|\ |: V \to \mathbb{R}$ satisfying

(1) $|x| \geq 0$ for all $x \in V$, and $|x| = 0$ if and only if $x = 0$.
(2) $|\alpha x| = |\alpha||x|$ for all $x \in V, \alpha \in \mathbb{R}$.
(3) $|x + y| \leq |x| + |y|$ for all pairs of vectors $x, y \in V$.

8.3. Continuity and Metric Spaces

A **normed linear space** is a vector space V with a norm defined on it.

EXAMPLE 8.3.2
$\mathbb{R}^n$ is an inner product space, where if $P_1 = (x_1, \ldots, x_n)$, $P_2 = (y_1, \ldots, y_n)$, then

$$< P_1, P_2 > = \sum_{i=1}^{n} x_i y_i.$$

Further, $\mathbb{R}^n$ is a normed linear space, where $|P_1| = (\sum_{i=1}^{n} x_i^2)^{1/2}$.

That these actually define an inner product and a norm on $\mathbb{R}^n$ is straightforward. However, $\mathbb{R}^n$ serves as a sort of generic example. Further, notice that if $P_1, P_2 \in \mathbb{R}^n$ then $d(P_1, P_2) = |P_1 - P_2|$ for the norm defined as above. □

These vector spaces are relevant to our discussion because of the following results.

Lemma 8.3.2
Any normed linear space V is a metric space, where $d(x,y) = |x - y|$ for $x, y \in V$.

Proof
Define $d(x, y) = |x - y|$. Then $d(x, y) \geq 0$ and $d(x, y) = 0$ if and only if $x = y$ follows directly from the norm property (1). Similarly, $d(x, y) = |x - y| = |-1||y - x| = d(y, x)$. Finally, $d(x, y) = |x - y| = |x - z + z - y| \leq |x - z| + |y - z| = d(x, z) + d(y, z)$. ∎

Lemma 8.3.3
Let V be an inner product space. For any $v \in V$ define $|v| = (< v, v >)^{1/2}$. This then defines a norm on V. In particular any inner product space is then a normed linear space and further then must be a metric space.

Proof
As in the proof of Lemma 8.3.2, this involves using the inner product properties to get the norm properties.

$|v| = (< v, v >)^{1/2}$, so $|v| \geq 0$ (since $< v, v > \geq 0$), and $|v| = 0$ if and only if $v = 0$.

$|\alpha v| = (< \alpha v, \alpha v >)^{1/2} = (\alpha^2 < v, v >)^{1/2} = |\alpha|(< v, v >)^{1/2} = |\alpha||v|$.

To get the final property we need the following important result, true in any inner product space, the proof of which is sketched in the exercises. ∎

Cauchy-Schwarz Inequality
In any inner product space V, for any $u,v \in V$,

$$|<u,v>| \leq |u||v|.$$

Now,

$$|u+v|^2 = <u+v, u+v> = <u,u> + 2<u,v> + <v,v>$$
$$\leq <u,u> + 2|<u,v>| + <v,v>$$
$$\leq |u|^2 + 2|u||v| + |v|^2 = (|u|+|v|)^2.$$

Taking squareroots we then get

$$|u+v| \leq |u|+|v|,$$

completing the proof.

Both $\mathbb{R}^n$ and the space $C^0[a,b]$ introduced in Example 8.3.1 are inner product spaces.

Theorem 8.3.1
Let $C^0[a,b]$ be the set of all continuous functions on the closed interval $[a,b]$. For $f(x), g(x) \in C^0[a,b]$ define

$$<f,g> = \int_a^b f(x)g(x)dx.$$

Then $C^0[a,b]$ is an inner product space and therefore a normed linear space and hence a metric space.

Proof
Addition, subtraction, and real scalar multiplication of continuous functions preserves continuity. Therefore, $C^0[a,b]$ is a real vector space. To show that it is an inner product space we must now show the inner product properties.

(1) $<f,f> = \int_a^b (f(x))^2 dx \geq 0$. If $<f,f> = 0$, then $\int_a^b (f(x))^2 dx = 0$, which implies that $f(x) \equiv 0$, since $f(x)$ is a continuous function.

(2) $<f,g> = \int_a^b f(x)g(x)dx = \int_a^b g(x)f(x)dx = <g,f>$.

(3) $<\alpha f, g> = \int_a^b \alpha f(x)g(x)dx = \alpha \int_a^b f(x)g(x)dx = \alpha <f,g>$.

(4) $<f+g, h> = \int_a^b (f(x)+g(x))h(x)dx$
$= \int_a^b f(x)h(x)dx + \int_a^b g(x)h(x)dx = <f,h> + <g,h>$. ∎

We now return to our main discussion. Since continuity relied only on distance, we can define continuity in any metric space.

Definition 8.3.5
Let $(M_1, d_1), (M_2, d_2)$ be metric spaces. A function $f : M_1 \to M_2$ is **continuous** at $x_0 \in M_1$ if for any $\epsilon > 0$ there exists a $\delta > 0$ such that

8.3. Continuity and Metric Spaces

$d(f(x), f(x_o)) < \epsilon$ whenever $d(x, x_o) < \delta$. The function f is **continuous** on M_1 if it is continuous at each point of M_1.

In order to extend these ideas to more general topological spaces, we must remove the dependence on the distance function. To accomplish this we introduce some other ideas.

Definition 8.3.6
Let (M, d) be a metric space. If $x_o \in M$, then an **open ball of radius** $\epsilon > 0$ centered on x_0, denoted by $S_\epsilon(x_o)$ is the set $\{x; d(x, x_0) < \epsilon\}$. A set $S \subset M$ is an **open set** if for all $x_0 \in S$ there exists an open ball centered on x_0 entirely contained in S. A set $C \subset M$ is a **closed set** if its complement C' is an open set.

Recall that these definitions generalize the concepts of open and closed regions in the complex plane as discussed in Chapter 4.

Theorem 8.3.2
Let (M,d) be a metric space. Then:

(1) Any union of open sets is open.
(2) Any intersection of finitely many open sets is open.

This theorem can be put succinctly by saying that in any metric space the class of open sets is closed under arbitrary unions and finite intersections.

We use these ideas to remove the distance function from the continuity definition. First of all, if $x \in S$ with S open, then S is called an **open neighborhood of** x. The next lemma is just a restatement of the continuity definition in terms of open balls.

Lemma 8.3.4
A function $f:M_1 \to M_2$ is continuous at $x_0 \in M_1$ if for each open ball $S_\epsilon(f(x_0))$ centered on $f(x_0)$ there exists an open ball $S_\delta(x_0)$ such that $f(S_\delta(x_0)) \subset S_\epsilon(f(x_0))$. Equivalently, for each open neighborhood S of $f(x_0)$, the pull-back $f^{-1}(S)$ is an open neighborhood of x_0.

The second part of the above lemma we can then extend.

Lemma 8.3.5
A function $f:M_1 \to M_2$ is continuous on M_1 if for each open set $O \subset M_2$, $f^{-1}(O)$ is an open set in M_1.

In the next section we will take Lemma 8.3.5 as the definition of continuity. Before continuing, we would like to go over some other ideas that the reader may be familiar with from analysis. These will not play a role in our final proof of the Fundamental Theorem of Algebra, but we want to tie them together with the above discussion.

We defined closed sets as being the complements of open sets. Most people first encounter closed sets as sets that contain their boundaries. These ideas are reconciled via the following concepts, which we just state.

Definition 8.3.7
Let (M, d) be a metric space and (x_n) a sequence of points in M. Then $x_n \to x_0$ if for any $\epsilon > 0$ there exists an N such that for all $n \geq N$, $x_n \in S_\epsilon(x_0)$. If $S \subset M$, then $x \in M$ is a **limit point** of S if $x_n \to x$ for some sequence in S. The point $x_0 \in S$ is an **interior point** if there exists an $\epsilon > 0$ such that $S_\epsilon(x_0) \subset S$. The **boundary** of S consists of all the limit points of S that are not interior points.

Theorem 8.3.3
Let (M,d) be a metric space. Then $C \subset M$ is closed if and only if C contains all its limit points.

Finally, in our definition of continuity the δ depended on both the given ϵ and the particular point x_0. In metric spaces we can strengthen this.

Definition 8.3.8
A function $f : M_1 \to M_2$ is **uniformly continuous** on M_1 if for any $\epsilon > 0$ there exists a $\delta > 0$, depending only on ϵ, such that $d(f(x_1), f(x_2)) < \epsilon$ whenever $d(x_1, x_2) < \delta$.

Clearly being uniformly continuous implies being continuous. In $\mathbb{R}^n$ the converse is true over nice enough domains.

Theorem 8.3.4
Let $f:\mathbb{R}^n \to \mathbb{R}^m$ be continuous on $M \subset \mathbb{R}^n$. If M is closed and bounded (compact), then f is uniformly continuous on M.

Uniform continuity is really a metric property. When we remove the metric, as we will do in the next section, we essentially lose the concept of uniform continuity.

8.4 Topological Spaces and Homeomorphisms

In the last section we discussed the idea of a continuous function from one metric space to another. The definition was formulated in terms of

8.4. Topological Spaces and Homeomorphisms

the metrics on each space but then reformulated in terms of open sets. What we wish to do now is remove the reliance on the metric but still allow continuity. To do this we must have a collection of open sets, and this leads us to the concept of a topological space.

Definition 8.4.1
Let X be a nonempty set. A **topology** on X is a class of subsets T satisfying

(1) The empty set $\emptyset$ and the whole set X are in T.
(2) The union of any class of subsets in T is in T.
(3) The intersection of any finite class of subsets in T is in T.

Thus a topology is a class of subsets, closed under arbitrary unions and finite intersections. A subset in T is called an **open set**, while the complement of an open set is a **closed set**.

A **topological space** is a set X together with a topology on it. The elements of X are referred to as **points**, and the topology is called the class of **open sets** of X.

In any metric space the class of opens sets is closed under arbitrary unions and finite intersections and therefore constitutes a topology.

Lemma 8.4.1
Any metric space is a topological space. In particular, $\mathbb{R}^n$ and more generally any inner product space, is a topological space.

Not every topological space is a metric space. We give some other examples.

EXAMPLE 8.4.1
Consider any nonempty set X, and let T be the class of all subsets of X. Then T is a topology on X, called the **discrete topology**. □

EXAMPLE 8.4.2
Let X be any nonempty set and let $T = \{\emptyset, X\}$. Then T is a topology on X. □

EXAMPLE 8.4.3
Let X be an infinite set and let T consist of the empty set $\emptyset$ and every subset whose complement is finite. Then T forms a topology (see the exercises), called the **cofinite topology** on X. □

An obvious question to ask, given so many diverse examples of topological spaces, is which are in fact metric spaces. A general topological space X is **metrizable** if there exists at least one metric on X whose open sets coincide with the given topology. In the next section we will state sufficient conditions for metrizability.

We can now extend the concept of continuity.

Definition 8.4.2
Suppose X, Y are topological spaces and $f : X \to Y$. Then f is a **continuous function**, or **continuous mapping**, if $f^{-1}(O)$ is an open set in X whenever O is an open set in Y. Therefore, continuous functions pull back open sets to open sets.

The function $f : X \to Y$ is an **open mapping** if $f(V)$ is an open set in Y whenever V is an open set in X. Therefore, open mappings preserve open sets.

Definition 8.4.3
A **homeomorphism** between two topological spaces X, Y is a one-to-one continuous function of X onto Y (thus a bijection) that is also an open mapping. Equivalently, we could define this as a continuous bijection whose inverse is also continuous.

If there exists a homeomorphism $f : X \to Y$ we say that X, Y are **homeomorphic spaces**.

Homeomorphism plays the role in topology that isomorphism played in algebra. That is homeomorphic spaces are essentially "topologically the same". The basic **classification problem** in topology is to classify (distinguish) topological spaces up to homeomorphism.

8.5 Some Further Properties of Topological Spaces

Before moving on to algebraic topology and our final proof of the Fundamental Theorem of Algebra we mention some further results and properties from point-set topology. In particular we consider four properties: **separation, metrizability, compactness,** and **connectedness**. These are all motivated by corresponding ideas in the Euclidean spaces $\mathbb{R}^n$.

Any metric space M has the following separation property. Consider two distinct points $x, y \in M$. Then there exist disjoint open sets O_1, O_2 with $x \in O_1, y \in O_2$. This is clear from Figure 8.3. We just draw open balls about x and y whose radii sum to less than the distance from x to y, which is nonzero, since x and y are distinct.

From this it is clear that any point in a metric space is itself a closed set. With a little work (see the exercises) we can extend this to the following property: Given a metric space M and two disjoint closed subsets F_1, F_2 in M, there exists disjoint open sets O_1, O_2 with $F_1 \subset O_1, F_2 \subset O_2$. We now

8.5. Some Further Properties of Topological Spaces

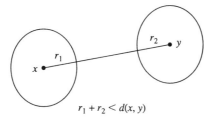

$r_1 + r_2 < d(x, y)$

Figure 8.3.

generalize these ideas to topological spaces, and from this we can then give sufficient conditions to ensure metrizability.

Definition 8.5.1
(1) A topological space X is a **T_1-space** if each point is a closed set.

(2) A topological space X is a **Hausdorff space** (or T_2-space) if given each pair x, y of distinct points in X there exist disjoint open sets O_x, O_y with $x \in O_x, y \in O_y$.

(3) A topological space X is a **completely regular space** (or T_3-space) if X is a T_1-space with the additional property that given a closed set F and a point $x \notin F$ there exists a continuous function $f : X \to [0, 1]$ such that $f(x) = 0$ and $f \equiv 1$ on F.

(4) A topological space X is a **normal space** (or T_4-space) if X is a T_1-space with the additional property that given disjoint closed subsets F_1, F_2 in X there exist disjoint open sets O_1, O_2 with $F_1 \subset O_1, F_2 \subset O_2$.

Theorem 8.5.1
For a topological space, normality implies complete regularity implies Hausdorff implies T_1.

From complete regularity on down the above theorem is straightforward. The fact that normality implies complete regularity is a consequence of a result called Urysohn's lemma (see the exercises). Further, from the discussion above we have that any metric space is normal and hence completely regular and Hausdorff.

Using the separation axioms, we can give criteria to guarantee metrizability. Recall from the last section that a topological space X is **metrizable** if there exists at least one metric on X whose topology coincides with the given topology on X.

Definition 8.5.2
If X is a topological space, an **open base** or **open basis** for X is a class B of open sets such that any open set is a union of sets from B. A **second-countable space** is a topological space with a countable open base.

Second countability together with normality are enough to ensure metrizability.

Theorem 8.5.2 (Urysohn Embedding Theorem)
If X is a second-countable normal space, then X is metrizable.

The proof of Theorem 8.5.2 is based on embedding (hence the name) such a space into the metric space $\mathbb{R}^\infty$. This is the space consisting of all sequences $\{x_1, \ldots, x_n, \ldots\}$ of real numbers such that $\sum_{i=1}^{\infty} |x_i|^2$ converges. On this space we define the metric as follows: if $x = \{x_1, \ldots, x_n, \ldots\}$, $y = \{y_1, \ldots, y_n, \ldots\}$, then $d(x,y) = (\sum_{i=1}^{\infty} |x_i - y_i|^2)^{1/2}$. This clearly generalizes the metric on $\mathbb{R}^n$ for finite n. If a space is embeddable in a metric space, it is itself a metric space.

The third property we consider is **compactness**. The Heine-Borel theorem in $\mathbb{R}^2$ states that if C is a closed and bounded subset of $\mathbb{R}^2$, then any covering of C by open sets has a finite subcovering. As a consequence of this theorem we obtain that continuous real-valued functions on closed and bounded domains are uniformly continuous. The Heine-Borel theorem and the statement on uniform continuity can be extended to any finite-dimensional real space $\mathbb{R}^n$. The general version of this property in a topological space is called **compactness**. In a topological space X a class of open sets $\{O_i\}$ is an **open cover** of X if $X \subset \cup_i O_i$. A **subcover** is a subclass that is also a cover.

Definition 8.5.3
A topological space X, or more generally a subset of X, is **compact** if every open cover has a finite subcover.

The Heine-Borel theorem can then be rephrased as:

Theorem 8.5.3 (Heine-Borel)
A subset of $\mathbb{R}^n$ is compact if and only if it is closed and bounded.

Compact spaces play an important role in topology. We mention two very nice results.

Theorem 8.5.4
Any compact Hausdorff space is normal.

Theorem 8.5.5
Let X be a compact space and $f:X \to Y$ a continuous function. Then the image of X is compact in Y.

Since subsets of $\mathbb{R}^n$ are compact if and only if they are closed and bounded, the above result implies that any continuous real-valued function on a compact set must be bounded. In particular, if $f(z)$ is a continuous complex function then the values $|f(z)|$ on any closed ball about the origin must be bounded.

The final property we look at is **connectedness**. Intuitively, a connected topological space is one that is in one piece. This property is closely tied to the intermediate value theorem, which has played such a large role in our proofs of the Fundamental Theorem of Algebra.

Definition 8.5.4
A topological space X, or more generally a subset Y of X, is **connected** if whenever $X = O_1 \cup O_2$ with O_1, O_2 open and $O_1 \cap O_2 = \emptyset$, then one of O_1 or O_2 must be empty: that is, X is connected if it cannot be decomposed as the union of two nonempty disjoint open sets. A decomposition into two nonempty disjoint open sets, when possible, is called a **disconnection**.

The ties to the intermediate value theorem are as follows.

Theorem 8.5.6
A subset $Y \subset \mathbb{R}$ is connected if and only if Y is an interval.

Theorem 8.5.7
If X is connected and $f:X \to Y$ is continuous, then $f(X)$ is connected.

Corollary 8.5.1
If $f:\mathbb{R} \to \mathbb{R}$ is continuous, then the image of an interval is an interval.

Notice that Corollary 8.5.1 is completely equivalent to the intermediate value theorem and can be generalized.

Corollary 8.5.2
Suppose $f:D \to \mathbb{R}$ is continuous, where D is a connected subset of $\mathbb{R}^n$. Then f assumes any value between any two of its values.

Exercises

8.1. Prove Lemma 8.3.1 – that is, distance in $\mathbb{R}^n$ satisfies all the basic metric properties.

8.2. Let $C^0[a, b]$ be the set of all continuous functions on the closed interval $[a, b]$. If $f(x), g(x) \in C^0[a, b]$, define

$$d(f(x), g(x)) = \left(\int_a^b |f(x) - g(x)|^2 dx \right)^{1/2}.$$

Prove that this defines a metric on $C^0[a, b]$ (Example 8.3.1), and so $C^0[a, b]$ equipped with this metric becomes a metric space.

8.3. Let $C^0[a, b]$ be the set of all continuous functions on the closed interval $[a, b]$. Now, if $f(x), g(x) \in C^0[a, b]$ define

$$d(f(x), g(x)) = \max_{[a,b]} |f(x) - g(x)|.$$

Prove that this also defines a metric on $C^0[a, b]$, and so again $C^0[a, b]$ becomes a (different) metric space with this metric.

8.4. Show that n-dimensional real space $\mathbb{R}^n$ forms an inner product space, where if $P_1 = (x_1, \ldots, x_n)$, $P_2 = (y_1, \ldots, y_n)$, then

$$< P_1, P_2 > = \sum_{i=1}^n x_i y_i$$

8.5. Prove the Cauchy-Schwarz inequality – that is, in any inner product space V, for any $u, v \in V$ then,

$$| < u, v > | \le |u||v|.$$

(Hint: Consider $< u - tv, u - tv >$ with $t \in \mathbb{R}$. Then let $t = \frac{<u,v>}{<v,v>}$.)

8.6. Prove Theorem 8.3.2 – that is, in a metric space (M, d)
(1) any union of open sets is open and
(2) any intersection of finitely many open sets is open.

8.7. Let (M, d) be a metric space. Then $C \subset M$ is closed if and only if C contains all its limit points (Theorem 8.3.3).

8.8. Let X be an infinite set and let T consist of the empty set $\emptyset$ and every subset whose complement is finite. Show that T then forms a topology (Example 8.4.3). T is called the **cofinite topology** on X.

8.9. Prove that given a metric space M and two disjoint closed subsets F_1, F_2 in M there exist disjoint open sets O_1, O_2 with $F_1 \subset O_1$, $F_2 \subset O_2$. (Hint: Prove first that in M if $x \notin F$ with F closed, then there exists an $\epsilon > 0$ such that $d(x, y) > \epsilon$ for all $y \in F$.)

8.10. Show that in any topological space complete regularity implies Hausdorff, which in turn implies T_1.

8.11. Urysohn's lemma says that in any normal space X, given disjoint closed subsets A, B there exists a continuous function $f : X \to [0, 1]$ with $f(A) \equiv 0$, $f(B) \equiv 1$. Use this to show that normality implies complete regularity.

8.12. Prove that in a Hausdorff space, given a point x and a compact subset F disjoint from it, then they can be separated by disjoint open sets. (Hint: For

Exercises

each $f \in F$ construct disjoint open sets O_x, O_f such that $x \in O_x, f \in O_f$. F is then contained in the union of the O_f, and then use the compactness.)

8.13. Prove that a compact subspace of a Hausdorff space is closed. (Hint: Use the previous exercise.)

8.14. Prove that a subset of $\mathbb{R}$ is connected if and only if it is an interval.

8.15. Show that the fact that the image of an interval in $\mathbb{R}$ under a continuous function is again an interval is equivalent to the intermediate value theorem.

8.16. Prove that the image of a compact set under a continuous map is compact (Theorem 8.5.5).

CHAPTER 9

Algebraic Topology and the Final Proof

9.1 Algebraic Topology

In the last chapter we gave our fifth proof of the Fundamental Theorem of Algebra by considering the winding numbers of polynomials. This led us to a discussion of topology and topological spaces. Now we present a more general proof modeled on the last one. To do this we must introduce and develop some of the basic ideas and techniques of **algebraic topology**.

Recall that topological spaces X, Y are homeomorphic if there exists a bijection $f : X \to Y$ such that both f and f^{-1} are continuous. The main problem in topology is to classify topological spaces up to homeomorphism. **Algebraic topology** is that branch of topology that attempts to solve this classification problem by assigning algebraic objects, primarily groups, to topological spaces. This assignment is done in such a manner that homeomorphic spaces correspond to isomorphic algebraic objects. Then if the corresponding algebraic objects can be shown to be nonisomorphic, the related topological spaces are nonhomeomorphic. The algebraic task is usually easier than the corresponding topological one, so by this change into an algebraic environment the classification problem is somewhat simplified.

Algebraic topology deals with finding **algebraic topological invariants**. A **topological invariant** is something (object, number, property) associated to a topological space that remains invariant (unchanged) under homeomorphism. An algebraic topological invariant is then an algebraic object associated to a topological space that remains invariant under homeomorphism.

9.1. Algebraic Topology

There are two general methods for finding groups associated to topological spaces – **homotopy theory** and **homology theory**. We discuss homotopy theory first.

If X, Y are topological spaces and $f : X \to Y, g : X \to Y$ are continuous maps, then roughly, f is **homotopic** to g if f can be continuously deformed into g, a notion that we will make precise in Section 9.3. The spaces X, Y are **homotopic**, or **homotopically equivalent** if there exist continuous maps $f : X \to Y, g : Y \to X$ such that $f \circ g$ is homotopic to the identity map on Y and $g \circ f$ is homotopic to the identity map on X. Very roughly, homotopy can be pictured geometrically as follows: X is homotopic to Y if X can be continuously deformed and then shrunk, stretched, or pinched to Y. If X, Y are homeomorphic they are clearly homotopic. However, spaces can be homotopic without being homeomorphic. For example the annular region pictured in Figure 9.1 is homotopic to the circle but not homeomorphic to it.

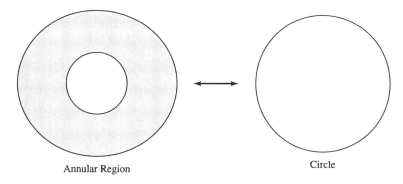

Annular Region Circle

Figure 9.1. Homotopic but not Homeomorphic

A **homotopy invariant** is a property of topological spaces preserved under homotopy. In Section 9.3 we introduce a group called the **fundamental group** that captures the homotopy properties of a space and that is a homotopy invariant.

Homology theory deals with topological spaces by considering them as combinatorial objects called **simplicial complexes**. Roughly, an n-**simplex** is an object similar to a polyhedron in n-dimensional real space. A **simplicial complex** is built up in a special way (see Section 9.4) from such simplices. If a general topological space can be built up from pieces that make it homeomorphic to a simplicial complex then this is called a **simplicial decomposition**, or **triangulation**, of the space. This combinatorial approach was initiated in pioneering work of Henri Poincare at the very end of the last century.

From the simplicial decomposition we can construct a sequence of abelian groups called the **homology groups**. Generally, homotopy is a stronger invariant then homology so that two spaces that are homotopically equivalent will have the same homology theory. We will make this more precise in Section 9.4.

9.2 Some Further Group Theory – Abelian Groups

The algebraic objects used in algebraic topology are primarily groups. Before proceeding further we must review some further group theory. In particular, we give a structure theorem for finitely generated abelian groups. This depends on the construction of direct products of groups.

Definition 9.2.1
Suppose H and K are groups. Then their **direct product**, denoted by $H \times K$, is the set $\{(h, k); h \in H, k \in K\}$ under the operation $(h_1, k_1)(h_2, k_2) = (h_1 h_2, k_1 k_2)$. It is straightforward to verify that $H \times K$ forms a group.

In the next lemma we summarize the results concerning direct products that are most relevant to our subsequent study. The proofs are straightforward and are left to the exercises.

Lemma 9.2.1
Suppose $G = H \times K$. Then:
  (1) $H_1 = H \times \{1\}, K_1 = \{1\} \times K$ are normal subgroups of G isomorphic to H, K respectively. Further, $G = H_1 K_1$ and $H_1 \cap K_1 = \{1\}$.
  (2) $H \times K$ is finite if and only if both H and K are finite.
  (3) $H \times K$ is abelian if and only if both H and K are abelian.

From the above lemma, part (1), we have that H and K can be considered as normal subgroups of $H \times K$. We can invert the process and start with subgroups of a given group G. If they satisfy the properties of Lemma 9.2.1, part (1), then G is the direct product of these subgroups.

Lemma 9.2.2
Suppose G is a group and H, K are normal subgroups of G satisfying
  (1) $G = HK$, that is, G is generated by all products $\{hk; h \in H, k \in K\}$,
  (2) $H \cap K = \{1\}$.
  Then $G \cong H \times K$. In this case we say that G is the direct product of its subgroups H, K.

We now turn our attention to abelian groups. Notice that from Lemma 9.2.1 direct products of abelian groups are themselves abelian. Recall that

9.2. Some Further Group Theory – Abelian Groups

a group G is **cyclic** if it has a single generator g. In this case G consists of all distinct powers of g. Clearly, cyclic groups are abelian, and therefore direct products of cyclic groups are also abelian. (They may or may not be cyclic – see the exercises.)

If a cyclic group has infinite order, it looks like the integers $\mathbb{Z}$ under addition. We will now use $\mathbb{Z}$ to denote an infinite cyclic group. If a cyclic group has finite order n, then it looks like the integers modulo n, $\mathbb{Z}_n$, under addition, and we will now use $\mathbb{Z}_n$ to denote a finite cyclic group of order n.

If G is any group and n is a natural number, by G^n we mean the direct product of n copies of G – that is, $G \times G \ldots \times G$. Thus $\mathbb{Z}^n$ is the direct product of n infinite cyclic groups. Direct products of infinite cyclic groups play an essential role in the structure of abelian groups and so we give them a special designation.

Definition 9.2.2
A **free abelian group** F **of rank** n is a direct product of n copies of $\mathbb{Z}$. That is,
$$F = \mathbb{Z} \times \ldots \times \mathbb{Z} = \mathbb{Z}^n.$$
If F is free abelian and $g_1, \ldots, g_n$ are elements of F such that
$$F = <g_1> \times <g_2> \times \ldots \times <g_n>,$$
then $g_1, \ldots, g_n$ are called a **basis** for F.

The terminology, basis and rank, is meant to mirror the corresponding terminology in vector spaces. Along these lines we have the following, which generalizes to free abelian groups several vector space properties.

Lemma 9.2.3
Let F be free abelian of rank n. Then:
 (1) The number of elements in any basis is unique and equal to n.
 (2) If $g_1, \ldots, g_n$ is a basis, then each $f \in F$ has a unique representation
$$f = g_1^{m_1} \ldots g_n^{m_n},$$
with each m_i an integer.
 (3) If H is a subgroup of F, then H is also free abelian, and rank $H \leq$ rank F.

The terminology **free** comes from the following categorical formulation of a free abelian group. This theorem essentially says that a free abelian group with basis $g_1, \ldots, g_n$ is the freest (least restrictive) abelian group that can be generated from $g_1, \ldots, g_n$.

Theorem 9.2.1
Let $g_1, \ldots, g_n$ be a set of elements in an abelian group G. Then G is free abelian with basis $g_1, \ldots, g_n$ if and only if each mapping $g_1 \to H, \ldots, g_n \to H$ into an abelian group H can be extended to a unique homomorphism of G to H.

An abelian group is **torsion-free** if it has no elements of finite order. It is a **torsion group** if every element has finite order. In an abelian group G all the elements of finite order form a subgroup, necessarily normal since G is abelian, called the **torsion subgroup**, denoted by tG.

Lemma 9.2.4
If G is abelian, all elements of finite order form a subgroup.

Proof
Let $tG = \{g \in G; g \text{ of finite order}\}$. Now, $1 \in tG$, so tG is nonempty. Suppose $g_1, g_2 \in tG$. Then there exist n_1, n_2 with $g_1^{n_1} = g_2^{n_2} = 1$. Then $(g_1 g_2)^{n_1 n_2} = (g_1^{n_1})^{n_2}(g_2^{n_2})^{n_1} = 1$, since G is abelian. Hence, $g_1 g_2$ has finite order and so $g_1 g_2 \in tG$. Similarly, $(g_1^{-1})^{n_1} = (g_1^{n_1})^{-1} = 1$, so $g_1^{-1} \in tG$. Therefore, tG is a subgroup. ∎

A group G is **finitely generated** if it has a finite system of generators. For abelian groups we get the following.

Lemma 9.2.5
If G is a finitely generated torsion abelian group then G is finite.

Proof
We will show this for two generators. The general case then follows easily by induction.

Suppose g_1, g_2 generate G. Since G is a torsion group, both g_1, g_2 have finite order, say n_1, n_2 respectively. Now, if $g \in G$, then $g = g_1^{t_1} g_2^{s_1} \cdots g_1^{t_k} g_2^{s_k}$, since g_1, g_2 generate G. Since G is abelian, we can put all the occurrences of g_1 together and all the occurrences of g_2 together so that

$$g = g_1^t g_2^s, \text{ for some integers } t, s.$$

Since this holds for all $g \in G$, the number of elements of G is bounded by the product of the number of choices for t and the number of choices for s. Since g_1 has order n_1 and g_2 has order n_2, this product is $n_1 n_2$. Therefore, $|G| \leq n_1 n_2$ and hence G is finite. ∎

We mention that the corresponding statement for nonabelian groups is not true. That is, there exist finitely generated groups where every element has finite order but that are infinite. The **Burnside problem** asks whether an n-generator group, where every element has finite order bounded by some other integer m, must be finite. In general this also has

9.2. Some Further Group Theory – Abelian Groups

a negative solution. We refer the reader to the survey article by Gupta [G] for a discussion of this.

Armed with these ideas we now state the **fundamental theorem for finitely generated abelian groups**. We outline its proof in a series of lemmas, however, we leave out most details. A very understandable complete discussion is in the book of Rotman [R]. After the outline of the proof, we will give an example of the use of this theorem.

Theorem 9.2.2
(*Fundamental Theorem for Finitely Generated Abelian Groups*) *Let G be a finitely generated abelian group. Then G has a unique (up to ordering of factors) decomposition as a direct product of a free abelian group of finite rank and finite cyclic groups of prime power order.*

The rank of the free abelian factor of G is an invariant of G called its **Betti number**.

As indicated, the following lemmas, 9.2.7 through 9.2.11, outline the proof of Theorem 9.2.2.

Lemma 9.2.7
If G is a finitely generated abelian group, then G is a direct product of cyclic groups.

Lemma 9.2.8
If G is a torsion-free finitely generated abelian group, then G is free abelian of finite rank.

Lemma 9.2.9
If G is a finitely generated abelian group, then $G = tG \times (\text{free abelian})$.

Lemma 9.2.10
If G is a finite abelian group, then G is a direct product of subgroups of prime power order.

If p is a prime, then a *p***-group** is a group where every element has order a power of p. Clearly, from Lagrange's Theorem if a finite group has prime power order then it is a p-group.

Lemma 9.2.11
If G is a finite abelian p-group, then G is a direct product of cyclic p-groups. Further, this decomposition is unique up to ordering of the factors.

We now give an example of the use of the theorem.

EXAMPLE 9.2.1
We classify all abelian groups with Betti number 3 and torsion group of order 180.

Suppose G is such an abelian group. Then

$$G \cong tG \times \text{(free abelian)}$$
$$= tG \times \mathbb{Z}^3,$$

since the Betti number is 3. Since $|tG| = 180$, we must then classify all finite abelian groups of order 180.

The prime factorization of 180 is $2^2 3^2 5$. Therefore, the possible group factorizations into cyclic groups of prime power order are:

(1) $\mathbb{Z}_{2^2} \times \mathbb{Z}_{3^2} \times \mathbb{Z}_5$.
(2) $\mathbb{Z}_2 \times \mathbb{Z}_2 \times \mathbb{Z}_{3^2} \times \mathbb{Z}_5$.
(3) $\mathbb{Z}_{2^2} \times \mathbb{Z}_3 \times \mathbb{Z}_3 \times \mathbb{Z}_5$.
(4) $\mathbb{Z}_2 \times \mathbb{Z}_2 \times \mathbb{Z}_3 \times \mathbb{Z}_3 \times \mathbb{Z}_5$.

Putting the two parts together, we see that there are precisely four possible abelian groups with the desired characteristics, namely:

(1) $\mathbb{Z}_{2^2} \times \mathbb{Z}_{3^2} \times \mathbb{Z}_5 \times \mathbb{Z}^3$.
(2) $\mathbb{Z}_2 \times \mathbb{Z}_2 \times \mathbb{Z}_{3^2} \times \mathbb{Z}_5 \times \mathbb{Z}^3$.
(3) $\mathbb{Z}_{2^2} \times \mathbb{Z}_3 \times \mathbb{Z}_3 \times \mathbb{Z}_5 \times \mathbb{Z}^3$.
(4) $\mathbb{Z}_2 \times \mathbb{Z}_2 \times \mathbb{Z}_3 \times \mathbb{Z}_3 \times \mathbb{Z}_5 \times \mathbb{Z}^3$. □

Before returning to topology we mention one more set of ideas from group theory. In an abelian group G the product $ghg^{-1}h^{-1}$ is always the identity for any elements $g, h \in G$. In general this is not true. If G is a non-abelian group, with $g, h \in G$ and $ghg^{-1}h^{-1} = 1$ then $gh = hg$ and we say that g and h **commute**. The collection of products of this sort in G describes how far from abelian the group is. We describe this more carefully.

Definition 9.2.3
Let G be a group. If $g, h \in G$, then their **commutator**, denoted $[g, h]$ is the product $ghg^{-1}h^{-1}$. Clearly g and h commute if and only if their commutator is trivial.

The **commutator subgroup** of G, denoted by $[G, G]$ or G', is the subgroup of G generated by all the commutators from G. Again, clearly G is an abelian group if and only if its commutator subgroup is trivial.

Theorem 9.2.3
Let G be a group. Then G' is a normal subgroup of G, and the quotient group G/G' is abelian. Further, if H is a normal subgroup of G with G/H abelian, then $G' \subset H$.

Hence, G/G' is the largest abelian quotient group of G. We call G/G' the **abelianization** of G, and denote this by G^{ab}.

9.3 Homotopy and the Fundamental Group

Proof
By definition G' is a subgroup. We must show that it is normal. Let $x \in G$ and consider

$$x^{-1}[g,h]x = x^{-1}ghg^{-1}h^{-1}x$$
$$= x^{-1}gxx^{-1}hxx^{-1}g^{-1}xx^{-1}h^{-1}x = [x^{-1}gx, x^{-1}hx] \in G'.$$

Therefore, every conjugate of a commutator is again a commutator, and hence G' is normal. ∎

Now consider the quotient G/G'. Suppose $u = gG', v = hG'$. Then $uvu^{-1}v^{-1} = ghg^{-1}h^{-1}G' = [g,h]G' = G'$ since $[g,h] \in G'$. But this implies that in G/G' the commutator $[u,v] = uvu^{-1}v^{-1} = 1$, and hence G/G' is abelian.

Finally, suppose G/H is abelian. Let $g, h \in G$ and consider the cosets gH, hH. Then

$$ghH = hgH,$$

since G/H is abelian. But then $ghg^{-1}h^{-1}H = H$, or $[g,h]H = H$. This implies that $[g,h] \in H$ for all commutators, and hence $G' \subset H$.

9.3 Homotopy and the Fundamental Group

We now introduce homotopy and from this build our first algebraic invariant, the fundamental group. Intuitively, if X, Y are topological spaces and $f : X \to Y, g : X \to Y$ are continuous maps then f is homotopic to g if it can be continuously deformed into g. We make this precise. For the remainder of this chapter we let $I = [0, 1]$ be the unit interval on the real line $\mathbb{R}$. Notice also that if X, Y are topological spaces, then their Cartesian product $X \times Y$ can also be made into a topological space by taking as a basis for the open sets in $X \times Y$ all sets of the form $O_x \times O_y$, where O_x, O_y are open in X, Y respectively (see the exercises).

Definition 9.3.1
Given topological spaces X, Y and continuous maps $f : X \to Y, g : X \to Y$, then f is **homotopic** to g if there exists a continuous map $H : X \times I \to Y$ such that $H(x, 0) = f(x)$ for all $x \in X$ and $H(x, 1) = g(x)$ for all $x \in X$. The mapping H is called a **homotopy**. In this case g is also homotopic to f via the homotopy $H_1(x, t) = H(x, 1 - t)$, so we say in general that f and g are **homotopic**.

To understand this definition consider the interval I as a **time parameter space**. By this we mean that at each point $t \in [0, 1]$ we consider the map $H_t(x) = H(x, t)$ as a continuous map from X to Y at time t. At time $t = 0$, $H_0(x) = f(x)$, while at time $t = 1$, $H_1(x) = g(x)$. The continuity of $H(x, t)$ represents the continuous deformation with time.

As an example, suppose $X = [a, b] \subset \mathbb{R}$ and $Y = \mathbb{R}^2$. Then f, g represent paths in $\mathbb{R}^2$. If they were homotopic, we would have a picture as in Figure 9.2. The dashed paths represent the paths at intermediate time values.

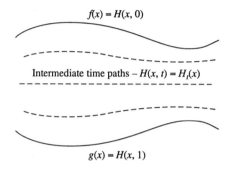

Figure 9.2. Homotopy of Paths and Maps

We have seen that if f is homotopic to g, then g is homotopic to f. Hence homotopy is a symmetric relation on continuous maps from X to Y. Further, it is a transitive relation. That is, if f is homotopic to g and g is homotopic to h, then f is homotopic to h. To see this, suppose H_0 is the homotopy from f to g and H_1 the homotopy from g to h. Then $H_2 : X \times I \to Y$ given by

$$H_2(x, t) = H_0(x, 2t), \text{ if } 0 \leq t \leq \frac{1}{2}$$

$$H_2(x, t) = H_1(x, 2t - 1), \text{ if } \frac{1}{2} \leq t \leq 1$$

gives the homotopy from f to h. Since clearly a map is homotopic to itself, the above comments show that homotopy is an equivalence relation on maps from X to Y. The equivalence classes are called **homotopy classes of maps**.

We now extend the concept of homotopy from maps to whole spaces.

Definition 9.3.2

Two topological spaces X, Y are **homotopically equivalent**, or **homotopic**, if there exist continuous maps $f : X \to Y$ and $g : Y \to X$ such that the composition gf is homotopic to the identity map on X and fg is homotopic to the identity map on Y.

9.3. Homotopy and the Fundamental Group

Very roughly, two spaces are homotopic if they can be continuously deformed and then pinched or shrunk into each other. Clearly homeomorphic spaces are homotopic, but there exist homotopic spaces that are not homeomorphic. An example was pictured in Figure 9.1. An annular region is homotopic to a circle but not homeomorphic to it.

Just as for maps, homotopy is an equivalence relation on topological spaces. Thus, we have **homotopy classes of spaces**. A **homotopic invariant** is a topological property preserved under homotopy. Thus, all spaces within a particular homotopy class have the same homotopy invariants.

A topological space is **contractible** if it is homotopic to a single point. For example, the interior of the unit disk in $\mathbb{R}^2$ is contractible. All contractible spaces are homotopic to one another.

We now construct the fundamental group of a space. This will be an algebraic homotopy invariant. A **path** in a general topological space X is a continuous map $\alpha : [0, 1] \to X$. If $x_0 \in X$, then a **loop**, or **closed path**, based at x_0 is a path α with $\alpha(0) = \alpha(1) = x_0$, that is, a path with x_0 as both its starting and ending point.

Let x_0 be a fixed point of X and let $C(X, x_0)$ denote the set of all loops based at x_0. We define an operation on the homotopy classes in $C(X, x_0)$ to form a group.

First of all, if α, β are loops at x_0, we define their product $\alpha\beta$ to be the loop that first goes around α (actually the image of α) and then goes around β. As a function we can express this as

$$\alpha\beta(x) = \alpha(2x), \text{ if } 0 \le t \le \frac{1}{2},$$

$$\alpha\beta(x) = \beta(2x - 1), \text{ if } \frac{1}{2} \le t \le 1.$$

We picture some of this in Figure 9.3.

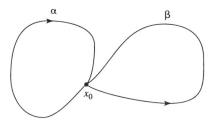

$\alpha\beta$ goes first around α and then around β

Figure 9.3. Product of Loops

This product of loops based at x_0 is preserved by homotopy. That is, if α is homotopic to α_1 and β is homotopic to β_1, then $\alpha\beta$ is homotopic to $\alpha_1\beta_1$. Further, this product is associative *up to homotopy*. This means that if α, β, γ are all loops based at x_0, then $(\alpha\beta)\gamma$ is homotopic to $\alpha(\beta\gamma)$. We summarize all these statements.

Lemma 9.3.1
Suppose α, β, γ are loops based at x_0. Then:
  (1) If α is homotopic to α_1 and β is homotopic to β_1, then $\alpha\beta$ is homotopic to $\alpha_1\beta_1$.
  (2) $(\alpha\beta)\gamma$ is homotopic to $\alpha(\beta\gamma)$.

If α is a loop at x_0, let $[\alpha]$ denote its homotopy class. Hence, $[\alpha]$ consists of all loops based at x_0 homotopic to α. Define $[\alpha][\beta] = [\alpha\beta]$; that is, the product of two homotopy classes of loops is the homotopy class of the product of class representatives. From Lemma 9.3.1 we can conclude that this is a well-defined operation on the set of homotopy classes. Further, from the second part of that lemma it follows that this is an associative operation on the set of homotopy classes. Let $\Pi_1(X, x_0)$ denote the set of homotopy classes of loops based at x_0. We now have a well-defined associative operation on $\Pi_1(X, x_0)$.

Let e denote the identity loop at x_0, that is, $e(t) = x_0$ for all $t \in [0, 1]$. Then $e\alpha = \alpha e = \alpha$ for any other loop based at x_0, and hence $[e][\alpha] = [\alpha][e] = [\alpha]$, and therefore, $[e]$ is an identity on $\Pi_1(X, x_0)$. Finally, if α is a loop at x_0, define $\alpha^{-1}(t) = \alpha(1 - t)$; that is, α^{-1} is the loop traversing α in the opposite direction. We can show that $\alpha\alpha^{-1}$ and $\alpha^{-1}\alpha$ are homotopic to e, and hence $[\alpha\alpha^{-1}] = [\alpha^{-1}\alpha] = [e]$. Therefore, each element of $\Pi_1(X, x_0)$ has an inverse. We thus have on $\Pi_1(X, x_0)$ a well-defined operation that is associative, has an identity, and under which there is an inverse for each element. It follows that $\Pi_1(X, x_0)$ forms a group.

Theorem 9.3.1
*The set $\Pi_1(X, x_0)$ of homotopy classes of loops based at $x_0 \in X$ forms a group under the operation $[\alpha][\beta] = [\alpha\beta]$. It is called the **fundamental group of** X **based at** x_0.*

The above definition depends on the base point x_0. We can in many cases remove this restriction. Suppose $x_1 \in X$ and suppose that there is a path p in X from x_0 to x_1, that is, a continuous function $p : [0, 1] \to X$ with $p(0) = x_0, p(1) = x_1$. Then $\Pi_1(X, x_0)$ is isomorphic to $\Pi_1(X, x_1)$ with the isomorphism being given by $[\alpha] \to [p^{-1}\alpha p]$. This is pictured in Figure 9.4.

Theorem 9.3.2
If x_0, x_1 are two points in the space X that are connected by a path p, then $\Pi_1(X, x_0) \cong \Pi_1(X, x_1)$.

9.3. Homotopy and the Fundamental Group

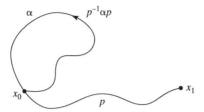

Figure 9.4. Isomorphism of $\Pi_1(X, x_0)$ and $\Pi_1(X, x_1)$

A topological space X is **path-connected** if any two points $x_0, x_1 \in X$ can be connected by a path. We then get the following important corollary to Theorem 9.3.2.

Corollary 9.3.1
*If X is path-connected, then $\Pi_1(X,x_0) \cong \Pi_1(X,x_1)$ for any pair of points $x_0, x_1 \in X$. It follows that in this case there is really only one fundamental group associated with X. We call this the **fundamental group** of X and denote it by $\Pi_1(X)$.*

For the remainder of this section we consider only path-connected spaces.

If the space X is contractible, then $\Pi_1(X) = \{1\}$. Further, if $\Pi_1(X) = \{1\}$ then every loop in X can be shrunk to a point. Such a space is called **simply connected**. We now give some examples.

EXAMPLE 9.3.1
(1) The unit disk in $\mathbb{R}^2$, or more generally any solid sphere in $\mathbb{R}^n$, is contractible and therefore simply connected.

(2) For $n > 1$ the n-sphere $S^n \subset \mathbb{R}^{n+1}$ is simply connected but not contractible. Recall that for $n \geq 0$ the n-sphere is $S^n = \{(x_1, \ldots, x_{n+1}) \in \mathbb{R}^{n+1}; x_1^2 + x_2^2 + \cdots + x_{n+1}^2 = 1\}$. □

EXAMPLE 9.3.2
Consider the circle $S^1 \subset \mathbb{R}^2$. Fix a base point on this circle and an orientation for paths. Any loop that does not go all the way around the circle is homotopic to a point. If a loop goes around a circle more than once but not twice, it is homotopic to a loop that goes around exactly once. Further, loops that go around exactly once are all homotopic. In general, if a loop is homotopic to a loop that goes around the circle exactly n times, we say that this loop has winding number n. All loops of the same winding number are homotopic.

With these ideas it follows that the homotopy classes of loops on the circle are precisely those loops with winding numbers $n = 0, 1, 2, \ldots$ and their inverses. The inverse of a loop with winding number n also winds n times, but in the opposite orientation. We say this has winding number

$(-n)$. Further, multiplication of loops is just addition of winding numbers. Thus, multiplication of homotopy classes here behaves like addition of integers, and hence we obtain that the fundamental group of the circle is infinite cyclic:

$$\Pi_1(S^1) = \mathbb{Z}.$$

□

EXAMPLE 9.3.3
Consider a torus T as in Figure 9.5.

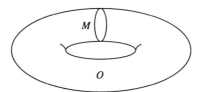

Figure 9.5. Torus

General loops are homotopic to loops that wind around the inside circle M and then the outside circle O. Such loops thus have a winding number around M and around O. Further, loops around M and around O commute. It follows that $\Pi_1(T)$ is free abelian of rank two, that is,

$$\Pi_1(T) = \mathbb{Z} \times \mathbb{Z}.$$

This result can also be obtained from the following. The torus T can be considered as a product space of two circles $T = S^1 \times S^1$. It can be proved that if X, Y are path-connected spaces, then

$$\Pi_1(X \times Y) = \Pi_1(X) \times \Pi_1(Y).$$

Therefore,

$$\Pi_1(T) = \Pi_1(S^1) \times \Pi_1(S^1) = \mathbb{Z} \times \mathbb{Z}.$$

□

Finally, an example to show that not every fundamental group is abelian.

EXAMPLE 9.3.4
Consider a figure eight curve in $\mathbb{R}^2$ as in Figure 9.6.

9.3. Homotopy and the Fundamental Group

Figure 9.6. Bouquet of Two Circles

As in the case of the torus, a loop is homotopic to a product of loops around one circle and then the other. However, since the curve is planar and one-dimensional, loops around the top circle do not commute with loops around the lower circle. Let α be a loop that winds once around the top circle and β a loop that winds once around the bottom circle with given orientations. Then a general loop is homotopic to a loop of the form

$$\alpha^{n_1}\beta^{m_1}\ldots\alpha^{n_k}\beta^{m_k} \text{ for integers } n_1, m_1, \ldots n_k, m_k.$$

Such a loop is called a **word** in α, β, and the corresponding group is called a **free group of rank 2.** This is nonabelian and not to be confused with a free abelian group of rank 2. We won't discuss these groups further here but refer to [Ro] for details.

More generally, a planar figure consisting of n circles touching at a single point is called a **bouquet of n circles** and the resulting fundamental group is a **free group of rank n.** □

The fundamental group is a homotopy invariant. In particular if X, Y are path-connected spaces and $h : X \to Y$ is a continuous map then for each loop α in X, $h(\alpha)$ is a loop in Y. Thus h defines a map $h^* : \Pi_1(X) \to \Pi_1(Y)$ by $h^*[\alpha] = [h(\alpha)]$. This map can be shown to be a group homomorphism.

Theorem 9.3.3
If X, Y are path-connected spaces and $h:X \to Y$ is a continuous map, then $h^*:\Pi_1(X) \to \Pi_1(Y)$ by $h^*[\alpha] = [h(\alpha)]$ is a homomorphism. If h is a homeomorphism, then h^* is an isomorphism. Further, if X, Y are homotopic then $\Pi_1(X)$ is also isomorphic to $\Pi_1(Y)$.

Example 9.3.5
Consider an annular region A as in Figure 9.1. Since A is homotopic to a circle, we have

$$\Pi_1(A) = \Pi_1(S^1) = \mathbb{Z}.$$

□

We mention that every group G can be considered as the fundamental group of a two-dimensional complex. Thus, in an abstract sense the study of group theory is really the study of fundamental groups. We refer the reader to either [Ro] or [C-Z] for a complete discussion of this.

We close this section by mentioning that there is a Galois theory associated to fundamental groups of spaces that is similar to the Galois theory of fields that we introduced in Chapter 7. We briefly describe this.

Definition 9.3.3
Let B, X be path-connected spaces. Then B is a **covering space** of X if there exists a map $p : B \to X$, called a **covering map**, such that for each $x \in X$ there exists an open neighborhood U of x such that each connected component of $p^{-1}(U)$ is homeomorphic to U under p.

Now, $p : B \to X$ defines a homomorphism $p^* : \Pi_1(B) \to \Pi_1(X)$. The covering map condition implies that this map is one-to-one, so $\Pi_1(B)$ can be considered as a subgroup of $\Pi_1(X)$. Now, if x is a particular base point in X then several (possibly infinite) points b_i may map onto x. The groups $\Pi_1(B, b_i)$ are all isomorphic and correspond to a complete set of conjugate subgroups in $\Pi_1(X)$. Conversely, given a complete set $\{H_i\}$ of conjugate subgroups in $\Pi_1(X)$, we can construct a path-connected covering space B with $\Pi_1(B) \cong H_i$. The simply connected covering space V of X corresponding to the trivial subgroup $\{1\}$ is called the **universal covering space** of X. We summarize these results, which are known as the Galois theory of covering spaces in Theorem 9.3.3.

Theorem 9.3.4
(1) Let B be a path-connected covering space of a path-connected space X with covering map p. Then:

(i) $p^*:\Pi_1(B) \to \Pi_1(X)$ is an injection so that $\Pi_1(B)$ can be considered as a subgroup of $\Pi_1(X)$.

(ii) If $x \in X$ and $p^{-1}(x) = \{b_i\} \subset B$, then $p^*(\Pi_1(B,b_i))$ forms a complete set of conjugate subgroups in $\Pi_1(X)$.

(2) Given a complete set $\{H_i\}$ of conjugate subgroups in $\Pi_1(X)$, there exists a path-connected covering space B of X such that $\Pi_1(B) \cong H_i$.

9.4 Homology Theory and Triangulations

In Euclidean n-space $\mathbb{R}^n$ consider the standard unit vectors $e_1 = (1, 0, \ldots, 0), e_2 = (0, 1, \ldots, 0), \ldots, e_n = (0, 0, \ldots, 1)$. Let $\Delta_0 = (0, 0, \ldots, 0)$ be the origin, and for $1 \leq q < n$ let Δ_q be the **convex hull**

9.4. Homology Theory and Triangulations

of $e_1, \ldots, e_{q+1}$. This is the set of vectors in $\mathbb{R}^n$ described by

$$\{u \in \mathbb{R}^n : u = a_1 e_1 + \cdots + a_{q+1} e_{q+1},$$

$$a_i \in \mathbb{R}, 0 \le a_i \le 1, 1 \le i \le q+1 \text{ and } \sum_{i=1}^{q+1} a_i = 1\}.$$

Thus Δ_0 is a point, Δ_1 a unit interval, Δ_2 a triangular region, Δ_3 a solid tetrahedron, and so on. A q-**dimensional simplex** in $\mathbb{R}^n$ is any subset homeomorphic to Δ_q. In a general topological space X a q-dimensional simplex is again any subset $D \subset X$ homeomorphic to Δ_q.

To build an algebraic invariant from these simplexes we must place an orientation on them. Now, let $P_0, P_1, \ldots, P_n$ be $n+1$ independent points in $\mathbb{R}^n$. By this we mean that the vectors $u_1 = \vec{P_0 P_1}, u_2 = \vec{P_0 P_2}, \ldots, u_n = \vec{P_0 P_n}$ are linearly independent vectors. An **oriented 0-simplex** is just a point and so is homeomorphic to Δ_0. An **oriented 1-simplex** is a directed line segment $P_0 P_1$ with the convention that $P_0 P_1 = -P_1 P_0$. This is equivalent to orienting Δ_1.

To define higher dimensional oriented simplexes we need some ideas about permutations. A permutation $\begin{pmatrix} 1 & 2 & \ldots & n \\ i_1 & i_1 & \ldots & i_n \end{pmatrix}$ is **even** if it can be decomposed into an even number of transpositions (see Exercise 7.18) or **odd** otherwise. This is equivalent to the sum of one less than each cycle length (again see Exercise 7.18) being even or odd. Hence $\begin{pmatrix} 1 & 2 & 3 \\ 2 & 3 & 1 \end{pmatrix} = (123)$ is even, but $\begin{pmatrix} 1 & 2 & 3 \\ 2 & 1 & 3 \end{pmatrix} = (12)$ is odd. Similarly, $\begin{pmatrix} 1 & 2 & 3 & 4 & 5 \\ 2 & 3 & 1 & 5 & 4 \end{pmatrix} = (123)(45)$ is odd since one less than the first cycle length is 2, one less than the second cycle length is 1, so the sum is 3.

Now, an **oriented 2-simplex** is a subset of $\mathbb{R}^n$ homeomorphic to Δ_2 and thus homeomorphic to a triangular region with a prescribed orientation on the three vertices. This is equivalent to giving a directed sequence of three points $P_0 P_1 P_2$. Now, $P_1 P_2 P_0$ gives the same orientation while $P_0 P_2 P_1$ gives the opposite orientation. In general $P_0 P_1 P_2 = P_i P_j P_k$ if the permutation $\begin{pmatrix} 1 & 2 & 3 \\ i & j & k \end{pmatrix}$ is even and $P_0 P_1 P_2 = -P_i P_j P_k$ if the permutation is odd. Hence

$$P_0 P_1 P_2 = P_1 P_2 P_0 = P_2 P_0 P_1 = -P_0 P_2 P_1 = P_1 P_0 P_2 = -P_2 P_1 P_0.$$

In general an **oriented q-simplex** is a subset of $\mathbb{R}^n$ homeomorphic to Δ_q with a prescribed orientation on its vertices. This is equivalent to designating the simplex as a directed sequence of points $P_0 P_1 \ldots P_q$ as above with $P_0 P_1 \ldots P_q = \pm P_{i_1} \ldots P_{i_q}$ depending on whether the corresponding permutation is even or odd. We will now designate an oriented q-simplex by a sequence of points such as $P_0 P_1 \ldots P_q$. For a general topological space an oriented q-simplex is a homeomorphic image of an oriented q-simplex in $\mathbb{R}^n$ with the orientation preserved.

For the q-simplex $P_0 P_1 \ldots P_q$ we use the notation $P_0 P_1 \ldots \overline{P_i} \ldots P_q$ to denote the $(q-1)$-simplex formed by omitting the point P_i. We then have:

Definition 9.4.1
Suppose $P_0 P_1 \ldots P_q$ is an oriented q-simplex. Then for $i = 0, \ldots, q$ its ith **face** is the oriented $(q-1)$-simplex given by

$$(-1)^i P_0 P_1 \ldots \overline{P_i} \ldots P_q.$$

Thus, a q-simplex has $(q+1)$ **faces**.

For example, the oriented 3-simplex $P_0 P_1 P_2 P_3$ has the 4 faces $P_1 P_2 P_3$, $-P_0 P_2 P_3$, $P_0 P_1 P_3$, $-P_0 P_1 P_2$.

For a 0-simplex we define its boundary ∂_0 to be the empty simplex, which is denoted by 0. Thus $\partial_0(P_0) = 0$. For $q \geq 1$ we define the **boundary** ∂_q of the q-simplex $P_0 P_1 \ldots P_q$ to be the formal sum of its faces. That is,

Definition 9.4.2
If $P_0 P_1 \ldots P_q$ is an oriented q-simplex with $q \geq 1$ then its **boundary** ∂_q is

$$\partial_q(P_0 P_1 \ldots P_q) = \sum_{i=1}^{q} (-1)^i P_0 P_1 \ldots \overline{P_i} \ldots P_q.$$

For the first four dimensions we have:

(1) $\partial_0(P_0) = 0$.
(2) $\partial_1(P_0 P_1) = P_1 - P_0$.
(3) $\partial_2(P_0 P_1 P_2) = P_1 P_2 - P_0 P_2 + P_0 P_1$.
(4) $\partial_3(P_0 P_1 P_2 P_3) = P_1 P_2 P_3 - P_0 P_2 P_3 + P_0 P_1 P_3 - P_0 P_1 P_2$.

In applying a boundary operator to a formal sum of faces we do it additively. Thus, for example,

$$\partial_1(P_1 P_2 - P_0 P_1) = \partial_1(P_1 P_2) - \partial_1(P_0 P_1)$$
$$= P_2 - P_1 - (P_1 - P_0) = P_2 - 2P_1 + P_0.$$

The following result is crucial to our construction of homology groups. It says that if we apply the boundary operator twice we always get 0.

Lemma 9.4.1
For any q-simplex we have $\partial_{q-1} \partial_q = 0$. Hence by additivity $\partial_{q-1} \partial_q$, applied to any formal sum of q-simplexes always gives 0.

9.4. Homology Theory and Triangulations

Proof

We show this for a 2-simplex. The computation for a 3-simplex is left to the exercises while the general case follows in the same manner.

Let $P_0P_1P_2$ be an oriented 2-simplex. Then

$$\begin{aligned}\partial_1\partial_2(P_0P_1P_2) &= \partial_1(P_1P_2 - P_0P_2 + P_0P_1) \\ &= \partial_1(P_1P_2) - \partial_1(P_0P_2) + \partial_1(P_0P_1) \\ &= P_2 - P_1 - P_2 + P_0 + P_1 - P_0 \\ &= 0.\end{aligned}$$

■

Many geometric figures in $\mathbb{R}^n$, in particular polyhedra, are built up in a very nice way from simplexes. These figures are called **simplicial complexes**, which we formally define below. First for a q-simplex we define its **subsimplexes** to be the set of all faces, all faces of faces, all faces of faces of faces, etc. without regard to orientation. Thus, for example, the set of subsimplexes of $P_0P_1P_2P_3$ is

$$P_1P_2P_3, P_0P_2P_3, P_0P_1P_3, P_0P_1P_2,$$
$$P_0P_1, P_0P_2, P_0P_3, P_1P_2, P_1P_3, P_2P_3, P_0, P_1, P_2.$$

Definition 9.4.3

A **simplicial complex** in $\mathbb{R}^n$ is a subset satisfying the following conditions:

(1) It is a union of oriented simplexes. Notice that in $\mathbb{R}^n$ the maximum dimension of a possible simplex is $n - 1$.

(2) Each point belongs to only finitely many simplexes.

(3) The intersection of two different (up to orientation) simplexes is either empty or one of the simplexes or a subsimplex of both.

In Figure 9.7 picture (A) is a simplicial complex in $\mathbb{R}^2$, while picture (B) is not since it violates (3).

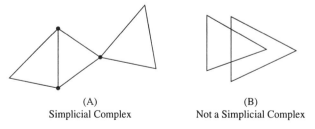

(A) Simplicial Complex

(B) Not a Simplicial Complex

Figure 9.7. Simplicial Complexes

We now extend this definition to general topological spaces. As indicated before, an oriented q-simplex in an arbitrary topological space X is a homeomorphism of an oriented q-simplex in some $\mathbb{R}^{n+1}$ with orientation preserved. Its faces and subsimplexes are carried into the corresponding faces and subsimplexes in X. Notice that in general there would be no bound on the dimension of a possible simplex.

Definition 9.4.3'
Let X be a topological space. Then a **simplicial complex** in X is a subset Y satisfying:

(1) It is a union of oriented simplexes. (In distinction to $\mathbb{R}^n$ there is no maximum dimension of a possible simplex.)

(2) Each point belongs to only finitely many simplexes.

(3) The intersection of two different (up to orientation) simplexes is either empty or one of the simplexes or a subsimplex of both.

If there is a maximum dimension q on the simplexes in Y, then Y is called a q-**dimensional simplicial complex**.

If X is a topological space that has a collection of subsets so that it can be built up as a simplicial complex, then we say that we have a **simplicial decomposition**, or **triangulation**, of X. It is from a simplicial decomposition of a space X that we will build our next algebraic invariant. First we give two examples in $\mathbb{R}^n$ of triangulations.

EXAMPLE 9.4.1
Consider S^1 the unit circle in $\mathbb{R}^2$. Then a triangulation as a 1-dimensional simplicial complex is pictured in Figure 9.8. The 0-simplexes are P_0, P_1, P_2, while the 1-simplexes are P_0P_1, P_1P_2, P_2P_0.

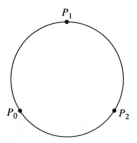

Figure 9.8. Circle Triangulation □

EXAMPLE 9.4.2
Let T be a torus. We can represent this in the plane by a rectangle with the sides identified, as in Figure 9.9. To recover the torus, fold the top side onto the identified bottom side to get a cylinder and then identify the top circle of the cylinder with the bottom circle.

9.4. Homology Theory and Triangulations

A triangulation of the torus is then pictured in Figure 9.10 where we triangulate the bounding circles as in the circle in Example 9.4.1 and then form the resulting triangles. □

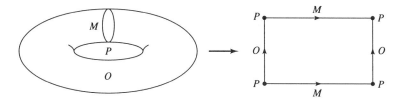

Figure 9.9. The Torus in the Plane

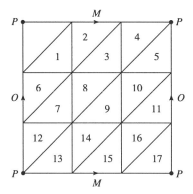

Figure 9.10. Triangulation of a Torus

Now we build groups from a triangulation. The spaces X we consider are simplicial complexes, and we also add the following finiteness condition: At each dimension q, there are only finitely many q-simplexes in X. This is not essential, but it is all we need for our applications and for the Fundamental Theorem proof. So now suppose X is a simplicial complex with the above condition. For each $n = 0, 1, 2, \ldots\ldots$ let $C_n(X)$ be the free abelian group of rank equal to the number of n-simplexes. We call this the n-**dimensional chain group**. An element of $C_n(X)$ is called an n-**dimensional chain**. Suppose X has t, n-simplexes $D_n^1, D_n^2, \ldots, D_n^t$. Then we can consider these as a basis for $C_n(X)$. Hence, an element of $C_n(X)$ can be considered as a sum

$$m_1 D_n^1 + m_2 D_n^2 + \ldots + m_t D_n^t, \text{ with } m_i \text{ arbitrary integers}. \quad (9.4.1)$$

(Note that since $C_n(X)$ is an abelian group, it is standard to employ additive notation, that is, to consider the group operation as addition.) Thus an n-dimensional chain is a sum of the form (9.4.1).

If D_n^k is an n-simplex in X, then its boundary $\partial_n D_n^k$, as defined earlier, is a sum of $(n-1)$-simplexes. Thus $\partial_n D_n^k \in C_{n-1}(X)$. This boundary operator,

∂_n, is defined on all elements of a basis, so it can be extended to a homomorphism of the whole group. Hence the **boundary homomorphism** is

$$\partial_n : C_n(X) \to C_{n-1}(X),$$

given by

$$\partial_n(m_1 D_n^1 + m_2 D_n^2 + \ldots + m_t D_n^t) = m_1 \partial_n(D_n^1) + \ldots + m_t \partial_n(D_n^t).$$

From Lemma 9.4.1 we have that $\partial_{n-1}\partial_n : C_n(X) \to C_{n-2}(X)$ is always the trivial homomorphism.

The kernel of ∂_n consists of all n-chains whose boundary is zero. We call these the n-**dimensional cycles**. The kernel is then a subgroup of $C_n(X)$, called the n-**dimensional cycle group**, denoted by $Z_n(X)$.

The image of $C_{n+1}(X)$ in $C_n(X)$ under ∂_{n+1} is called the n-**dimensional boundary group**, denoted by $B_n(X)$. Thus, an n-chain is in $B_n(X)$ if it is the boundary of an $(n+1)$-chain. From the fact that $\partial_n \partial_{n+1} = 0$ we have that $B_n(X)$ must be a subgroup of $Z_n(X)$, that is, each n-dimensional boundary is an n-dimensional cycle.

Since all these groups are abelian, all subgroups are normal, and hence $B_n(X)$ is a normal subgroup of $Z_n(X)$. Therefore, we can form the quotient group $Z_n(X)/B_n(X)$. This quotient group is called the n-**dimensional homology group**, denoted by $H_n(X)$. We summarize all this.

Definition 9.4.4
If X is a simplicial complex with the above condition, then:

(1) $C_n(X)$ is the n-**dimensional chain group**, which is the free abelian group with basis the n-simplexes in X.

(2) $\partial_n : C_n(X) \to C_{n-1}(X)$ is the **boundary operator**, which is the homomorphism formed by extending the boundary operator on simplexes.

(3) $Z_n(X)$ is the n-**dimensional cycle group**, which is the kernel of the boundary homomorphism $\partial_n : C_n(X) \to C_{n-1}(X)$.

(4) $B_n(X)$ is the n-**dimensional boundary group**, which is the image of $C_{n+1}(X)$ in $C_n(X)$ under the boundary homomorphism ∂_{n+1}.

(5) $H_n(X)$ is the n-**dimensional homology group**, which is $Z_n(X)/B_n(X)$.

Two cycles in $C_n(X)$ that differ by a boundary – that is, lie in the same coset of $B_n(X)$, are said to be **homologous**.

These definitions appear to depend on the particular triangulation of the space X. That is, if X had a different simplicial complex then perhaps the homology groups might be different. The following important **invariance** theorem says that this cannot be true and is really where the strength of the combinatorial topological approach lies.

Theorem 9.4.1
Suppose X is a topological space that has a simplicial decomposition. Then the homology groups determined from this decomposition are isomorphic to those determined by any other simplicial decomposition of X.

Homology is also a topological invariant, actually a homotopy invariant. Each continuous map $f : X \to Y$ defines a map on simplexes and can be extended to a homomorphism between homology groups. This is similar to what we did for the fundamental group in the last section. If this continuous map is a homeomorphism then the resulting homomorphism is an isomorphism.

Theorem 9.4.2
Suppose X, Y are topological spaces with simplicial decompositions and $h: X \to Y$ is a continuous map. Then h induces a homomorphism $h^*: H_n(X) \to H_n(Y)$ for all n. If h is a homeomorphism, then h^* is an isomorphism. Further, if X, Y are homotopic then the homology groups are also isomorphic.

The last statement in Theorem 9.4.2 says that homotopy is a stronger invariant than homology, that is, homotopic spaces cannot be distinguished by homology.

9.5 Some Homology Computations

Before moving on to the homology of general spheres and Brouwer degree, which are the basis for our final proof of the Fundamental Theorem of Algebra, we give some elementary examples of homology calculations. To find the homology of a space X we need a simplicial decomposition – any one will do, from Theorem 9.4.1. This is in most cases quite difficult, so consequently many techniques and results have been developed to aid in the computation of homology. We mention several of these that we will use in our examples.

If X has a q-dimensional simplicial decomposition, that is, no simplexes of dimension higher than q, then clearly $H_n(X) = 0$ if $n > q$.

Next, if X is a contractible space, then it is homotopic to a point. Since homotopy preserves homology, the homology of a contractible space is the same as the homology of a point P. Clearly, $Z_0(P) = \mathbb{Z}$ while $B_0(P) = 0$ since P is 0-dimensional. Therefore, $H_0(P) = \mathbb{Z}$, and $H_n(P) = 0$ if $n \geq 1$.

Theorem 9.5.1
If X is a contractible space, then $H_0(X) = \mathbb{Z}$ and $H_n(X) = 0$ if $n \geq 1$.

Now, if the space X is not connected, it will decompose into a disjoint union of connected pieces. These are called the **connected components** of X. Suppose X has a simplicial decomposition. Then within each of these connected components each vertex point P is a cycle. Further, if $P_0, P_1, \ldots, P_t$ are the vertices within a component, then

$$P_k = P_0 + (P_1 - P_0) + \cdots + (P_k - P_{k-1}).$$

Now, each $P_i - P_j$ is a boundary and hence in $B_0(X)$. Therefore, each vertex is in the same coset of $B_0(X)$. It follows that $H_0(X) = Z_0(X)/B_0(X)$ is just free abelian of rank one with a basis of a single vertex. The same would be true within each other connected component. Therefore, we have the following result.

Theorem 9.5.2
If X is a simplicial complex with n connected components, then

$$H_0(X) = \text{free abelian of rank } n \cong \mathbb{Z}^n.$$

Finally, the following result can be proved. This not only allows a computation of H_1 when Π_1 is known but gives a relationship between homology and homotopy. Recall that if G is a group, its abelianization is $G^{ab} = G/G'$ (see Section 9.2). If G is abelian, then $G^{ab} = G$.

Theorem 9.5.3
If X is a path-connected simplicial complex then

$$H_1(X) \cong \Pi_1(X)^{ab}.$$

If $\Pi_1(X)$ is abelian then $H_1(X) \cong \Pi_1(X)$.

EXAMPLE 9.5.1
We compute the homology of the circle S^1. A triangulation was given in Figure 9.8. Clearly, the homology of S^1 is the same as the homology of a triangle, the boundary of a 2-simplex.

Since S^1 is connected it has one connected component and we have $H_0(S^1) = \mathbb{Z}$. Further, since S^1 is 1-dimensional, $H_n(S^1) = 0$ for $n > 1$.

We saw in Section 9.3 that $\Pi_1(S^1) = \mathbb{Z}$ which is abelian. Therefore, $H_1(S^1) \cong \Pi_1(S^1) = \mathbb{Z}$. We could also obtain this result directly from the triangulation.

$C_1(S^1)$ is free abelian with basis P_0P_1, P_1P_2, P_2P_0. A 1-chain is then of the form

$$m_1 P_0 P_1 + m_2 P_1 P_2 + m_3 P_2 P_0 \text{ for integers } m_1, m_2, m_3.$$

The only way the vertex P_1 will be eliminated in forming the boundary is if $m_1 = -m_2$. Similarly, the only way P_2 will be eliminated is if $m_3 = -m_2$.

9.5. Some Homology Computations

It follows that a 1-cycle must have the form

$$m(P_0P_1 - P_1P_2 + P_2P_0), m \in \mathbb{Z}.$$

Thus, any cycle must be a multiple of the single 1-chain $P_0P_1 - P_1P_2 + P_2P_0$. It follows that $Z_1(S^1) \cong \mathbb{Z}$. Now, $B_1(S^1) = 0$, since there are no 2-chains, so $H_1(S^1) = Z_1(S^1)/B_1(S^1) = \mathbb{Z}$. □

EXAMPLE 9.5.2
Consider the 2-sphere S^2. Just as S^1 could be triangulated to look like a triangle, S^2 can be triangulated to look like the surface of a tetrahedron, as pictured in Figure 9.11.

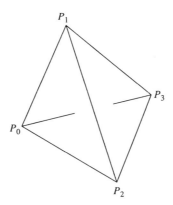

Figure 9.11. Triangulation of the 2-Sphere

Now, this is connected, so $H_0(S^2) = \mathbb{Z}$. Any loop on the surface of a sphere can clearly be shrunk to a point, so that $\Pi_1(S^2) = 0$. From Theorem 9.5.3, then, $H_1(S^2) = 0$. Since this is 2-dimensional, we have $H_n(S^2) = 0$ if $n > 2$. We now consider $H_2(S^2)$.

A 2-chain has the form

$$m_1P_0P_1P_2 + m_2P_1P_2P_3 + m_3P_2P_3P_0 + m_4P_3P_0P_1, m_i \in \mathbb{Z}.$$

If this were a cycle, the common edge of the faces $P_0P_1P_2$, $P_1P_2P_3$, which is P_1P_2, would have to be eliminated. This would imply that $m_2 = -m_1$. By an identical argument on the other faces we then have that a cycle must have the form

$$m(P_0P_1P_2 - P_1P_2P_3 + P_2P_3P_0 - P_3P_0P_1), m \in \mathbb{Z},$$

that is, a multiple of a single 2-chain. It follows that $Z_2(S^2) = \mathbb{Z}$. We have $B_2(S^2) = 0$, since it is 2-dimensional and there are no 3-chains. Then $H_2(S^2) = Z_2(S^2)/B_2(S^2) = \mathbb{Z}$.

Hence in summary,

$$H_0(S^2); = \mathbb{Z};$$
$$H_2(S^2) = \mathbb{Z};$$
$$H_n(S^2) = 0 \text{ if } n \neq 0, 2.$$

□

EXAMPLE 9.5.3
Consider a torus T. A triangulation was pictured in Figure 9.10. T is connected, so $H_0(T) = \mathbb{Z}$. The group $\Pi_1(T)$ was free abelian of rank 2, so $H_1(T) \cong \Pi_1(T) = \mathbb{Z} \times \mathbb{Z}$.

A 2-chain is a sum of the seventeen triangles (2-simplexes) pictured in Figure 9.10. As in the argument for the circle and in the argument for the sphere, the only way the common edge of triangles numbered 1,2 will cancel in taking the boundary is if the coefficients are of the same magnitude but of different signs. If we let $T_1, \ldots, T_{17}$, denote these 2-simplexes then this argument leads to the fact that a 2-cycle in T is a multiple of $T_1 - T_2 + T_3 - \cdots + T_{17}$. It follows that $Z_2(T) = \mathbb{Z}$, and since $B_2(T) = 0$, the torus being 2-dimensional, we have $H_2(T) = \mathbb{Z}$.

Finally, $H_n(T) = 0$ if $n > 2$. □

EXAMPLE 9.5.4
Consider X to be a figure eight – bouquet of two circles as seen in Figure 9.6. If we triangulate each circle as in Example 9.5.1 we get a triangulation of this figure as a 1-dimensional simplicial complex.

Since this figure is connected $H_0(X) = \mathbb{Z}$. The group $\Pi_1(X)$ is a free group of rank 2 (see Section 9.3). The abelianization of such a group is free abelian of rank 2. Therefore, $H_1(X) = \Pi_1(X)^{ab} = \mathbb{Z} \times \mathbb{Z}$. Finally, $H_n(X) = 0$ if $n > 1$. □

9.6 Homology of Spheres and Brouwer Degree

We saw in Example 9.5.2 that if $n \geq 1$, then $H_n(S^2) = 0$ if $n \neq 2$, and $H_2(S^2) = \mathbb{Z}$. This is a general property of spheres that is crucial in forming our final proof. Recall that an n-sphere is a set homeomorphic to $S^n = \{(x_0, \ldots, x_n) \in \mathbb{R}^{n+1}; x_0^2 + \cdots + x_n^2 = 1\}$. In general, an n-sphere is homeomorphic to the surface of an $(n+1)$-simplex. Thus, the 1-sphere S^1 is homeomorphic to a triangle, the 2-sphere S^2 is homeomorphic to the surface of a tetrahedron, and so on. By extending and refining the

9.6. Homology of Spheres and Brouwer Degree

arguments used in Examples 9.5.1 and 9.5.2, the following can be proved. We refer to [Be] for a detailed proof.

Theorem 9.6.1
Let S^n be an n-sphere with $n \geq 1$. Then $H_0(S^n) = \mathbb{Z}$, and for $q \geq 1$,

$$H_n(S^n) = \mathbb{Z},$$
$$H_q(S^n) = 0 \text{ if } n \neq q.$$

Now, suppose S^n, Σ^n are two n-spheres, and consider a continuous map $f : S^n \to \Sigma^n$. This induces a homomorphism $f^* : H_n(S^n) \to H_n(\Sigma^n)$. From Theorem 9.6.1, $H_n(S^n)$ is infinite cyclic. Suppose α is a generator of $H_n(S^n)$. Similarly, $H_n(\Sigma^n)$ is infinite cyclic and suppose β is a generator. Then since f^* is a homomorphism, $f^*(\alpha)$ is an integer multiple of β. That is,

$$f^*(\alpha) = m\beta \text{ for some integer } m.$$

The number m is called the **Brouwer degree**, or **degree**, of the map f. We abbreviate this by $deg(f)$. Intuitively, $deg(f)$ is the algebraic number of times that the image $f(S^n)$ winds around Σ^n.

The following three results on degree are the critical ones for the proof of the Fundamental Theorem of Algebra. The proofs are lengthy and we leave them out. They can be found in [H-Y].

Lemma 9.6.1
The degree of a continuous map is independent of the simplicial decompositions of S^n and Σ^n.

This lemma is needed because even though $H_n(S^n)$ and $H_n(\Sigma^n)$ are both infinite cyclic, the action of f on the cycle α generating $H_n(S^n)$ may be different for different simplicial decompositions. The lemma says that in mapping to another sphere, up to homology this does not happen.

Lemma 9.6.2
(1) The degree of a continuous map f of an n-sphere S^n into an n-sphere Σ^n depends only on the homotopy class of f. In particular, two homotopic maps $f:S^n \to \Sigma^n, g:S^n \to \Sigma^n$ have the same degree.

(2) If $f:S^n \to \Sigma^n, g:S^n \to \Sigma^n$ are two continuous maps with $deg(f) = deg(g)$, then f and g are homotopic.

This lemma serves to classify completely, up to homotopy, the continuous mappings between n-spheres. In particular, the homotopy classes are in one-to-one correspondence with the integers.

Theorem 9.6.2
If $f:S^n \to \Sigma^n$ and $\deg(f) \neq 0$, then each point of Σ^n is in the image of $f(S^n)$.

This last theorem will be the key result in our final proof.

Before closing this section, we state a famous result due to Brouwer – the **Brouwer fixed point theorem** – that also is proved using Brouwer degree. If $f : X \to X$, then a **fixed point** for f is a point $x_0 \in X$ such that $f(x_0) = x_0$.

Theorem 9.6.3
(Brouwer Fixed Point Theorem) Let D^n be the n-ball, that is, the interior and boundary of the unit sphere $S^{n-1} \subset \mathbb{R}^n$. Suppose $f:D^n \to D^n$ is a continuous map. Then f has at least one fixed point.

9.7 The Fundamental Theorem of Algebra – Proof Six

We can now give our final proof. Consider a unit sphere S^2 tangent to the complex plane $\mathbb{C}$ at the origin $(0, 0, 0)$ as in Figure 9.12.

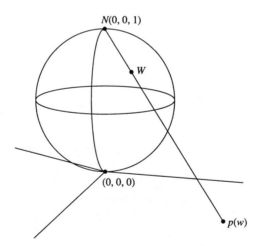

Figure 9.12. Stereographic Projection

We map this sphere onto $\mathbb{C}$ in the following manner. Let $N = (0, 0, 1)$ be the north pole, and for any point $w \in S^2 - N$ form the line joining w to N. This line will intersect $\mathbb{C}$ in a unique point $p(w)$. This mapping is called **stereographic projection**. The equations for this map can be

9.7. The Fundamental Theorem of Algebra – Proof Six

explicitly derived (see [A]), but all we need is that p is a continuous bijection between $\mathbb{C}$ and $S^2 - N$. If we identify the north pole N with ∞, then p is a bijection of all of S^2 with $\mathbb{C} \cup \infty$. The map p can be shown to be conformal, and thus S^2 provides a conformal model of the extended complex plane $\mathbb{C} \cup \infty$. When we look at S^2 in this manner, it is called the **Riemann Sphere**.

Theorem 9.7.1
Stereographic projection p provides a conformal bijection between S^2 and $\mathbb{C} \cup \infty$.

Now, suppose $P(z) = a_0 + a_1 z + \cdots + a_n z^n$, $a_n \neq 0$, is a nonconstant complex polynomial of degree n. Since we are trying to find a root of $P(z)$ and $a_n \neq 0$, we can without loss of generality assume that $a_n = 1$. Therefore, consider $P(z) = a_0 + a_1 z + \cdots + z^n$, $n \geq 1$. Since $P(\infty) = \infty$, $P(z)$ can then be considered as mapping the Riemann sphere to the Riemann sphere, that is, $P(z)$ is a continuous map $S^2 \to S^2$. As such it has a Brouwer degree m. We first show that its Brouwer degree is the same as its polynomial degree.

Lemma 9.7.1
$P(z)$ is homotopic to the map $f(z) = z^n$.

Proof
Let $H(z, t) = z^n + (1 - t)(a_0 + a_1 z + \cdots + a_{n-1} z^{n-1})$ for $z \in \mathbb{C}, t \in [0, 1]$ and $H(\infty, t) = \infty$ for $t \in [0, 1]$. Clearly, $H(z, t)$ is continuous on $\mathbb{C} \times I \to \mathbb{C}$. Now, $\lim_{z \to \infty} H(z, t) = \infty = H(\infty, t)$ for all $t \in [0, 1]$, so $H(z, t)$ is continuous from $(\mathbb{C} \cup \infty) \times I \to \mathbb{C} \cup \infty$, and thus from $S^2 \times I \to S^2$. Since $H(z, 0) = P(z)$ and $H(z, 1) = z^n$, this defines a homotopy. ∎

Lemma 9.7.2
The Brouwer degree of $f(z) = z^n$ is n. Thus for complex polynomials, Brouwer degree and polynomial degree coincide.

Proof
Consider the two triangulations of S^2 (considered as $\mathbb{C} \cup \infty$) pictured in Figure 9.13.

Under $f(z) = z^n$ the n-shaded pieces in figure (A) are mapped onto the one-shaded piece in figure (B) in an orientation-preserving fashion. It is clear then that f will map a cycle in the triangulation (A) to n times the cycle in triangulation (B). Therefore, the degree of z^n is n. ∎

Theorem 9.7.2
(Fundamental Theorem of Algebra) Suppose $P(z)$ is a nonconstant complex polynomial. Then $P(z)$ has a complex root.

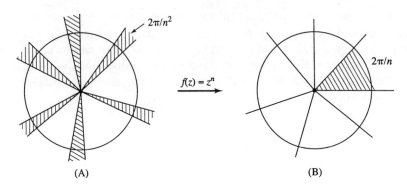

Figure 9.13. Winding of $f(z) = z^n$

Proof

Suppose $P(z) = a_0 + a_1 z + \cdots + a_n z^n$. Since $P(z)$ is assumed nonconstant, $n > 0$ and $a_n \neq 0$. Therefore, as before we can assume that $a_n = 1$. From Lemma 9.7.1, $P(z)$ is homotopic to $f(z) = z^n$, and therefore, from Lemma 9.7.2 and Lemma 9.6.2, $P(z)$ has Brouwer degree $n \neq 0$. From Theorem 9.6.2 each point of S^2 is then in the image of $P(z)$. In particular there exists at least one point $z_0 \in S^2$ with $P(z_0) = 0$. Since $P(\infty) = \infty$, it follows that $z_0 \in \mathbb{C}$. ∎

9.8 Concluding Remarks

This completes our final topological proof and the last of the three pairs of proofs toward which these notes were directed. Notice that in the topological situation both the straightforward proof given in Section 8.1 and the final proof given in the last section involved the use of winding number. Notice also that for the Fundamental Theorem of Algebra we really didn't need the full power of general Brouwer degree, only Brouwer degree for 2-spheres. In Appendix D we present two additional topologically motivated proofs that are slightly different variations on this theme.

Exercises

9.1. Prove that if H and K are groups, then their Cartesian product $H \times K$ is also a group under $(h, k)(h_1, k_1) = (hh_1, kk_1)$.

9.2. Prove Lemma 9.2.1.

9.3. Prove Lemma 9.2.2.

Exercises

9.4. Prove that if p, q are relatively prime integers, then $\mathbb{Z}_p \times \mathbb{Z}_q$ is cyclic. This shows that the direct product of cyclic groups can be cyclic.

9.5. Show that $\mathbb{Z}_2 \times \mathbb{Z}_2$ is not cyclic. This shows that the direct product of cyclic groups may not be cyclic.

9.6. Prove that every finitely generated abelian group is a direct product of cyclic groups. (Hint: Show it for two generators.)

9.7. Prove that a finite abelian group is a direct product of abelian p-groups. (Hint: In G consider all elements of order a power of p for p a prime dividing the order of G. Show that this is a subgroup and G is the direct product of all the subgroups of this form for all primes dividing the order of G.)

9.8. Give a complete classification of all finite abelian groups of the following orders. (i) 25 (ii) 45 (iii) 37 (iv) 1284

9.9. Show that if X, Y are homeomorphic, then they are homotopic.

9.10. Show that all contractible spaces are homotopic.

9.11. Describe a general element of the fundamental group of a bouquet of n circles.

9.12. Suppose $\Pi_1(X) = \mathbb{Z}_2 \times \mathbb{Z}_2$. Up to homotopy, how many possible covering spaces of X can there be?

9.13. Determine the homology of a bouquet of n circles.

9.14. If $P_0P_1P_2P_3$ is a 3-simplex, verify that $\partial_2\partial_3(P_0P_1P_2P_3) = 0$.

9.15. Compute the homology of two tangent 2-spheres.

APPENDIX A

A Version of Gauss's Original Proof

Gauss's first proof of the Fundamental Theorem of Algebra was essentially different from any of the six that we have looked at up to this point. This original proof was given in his Ph.D. dissertation of 1799, and his fourth proof, published in 1849, is another presentation of the first. In this section we give a relatively modern version of Gauss's original proof. This demonstration borrows freely from the book of Uspensky [U].

Suppose

$$f(z) = z^n + a_{n-1}z^{n-1} + \cdots + a_0$$

is a nonconstant complex polynomial. As before, we can assume that the leading coefficient is 1 and the constant term a_0 is nonzero. Further, we can assume that $n > 2$. The case $n = 1$ is just a linear polynomial and thus clearly has a root, while the case $n = 2$ is handled by the quadratic formula. With the assumption that $n > 2$ suppose the coefficients are given in polar form by

$$a_{n-1} = A(\cos\alpha + i\sin\alpha), a_{n-2} = B(\cos\beta + i\sin\beta), \ldots,$$

$$\ldots, a_0 = L(\cos\lambda + i\sin\lambda)$$

and the variable $z = r(\cos\phi + i\sin\phi)$. The polynomial $f(z)$ can be written in terms of its real and complex parts (see Chapter 4) as

$$f(z) = T(z) + iU(z). \qquad (A.1)$$

Using DeMoivre's theorem and the polar representation of the coefficients, T and U can be expressed explicitly in terms of r and ϕ as

$$T = r^n \cos n\phi + Ar^{n-1}\cos((n-1)\phi + \alpha) + \cdots + L\cos\lambda, \qquad (A.2)$$

Appendix A: A Version of Gauss's Original Proof

$$U = r^n \sin n\phi + Ar^{n-1} \sin((n-1)\phi + \alpha) + \cdots + L \sin \lambda. \quad \text{(A.3)}$$

To prove the Fundamental Theorem of Algebra we must show that there exists a point z_0 where $T(z_0)$, $U(z_0)$ are simultaneously zero.

First of all, it is possible to find a real number R such that for $r > R$,

$$r^n - \sqrt{2}(Ar^{n-1} + Br^{n-2} + \cdots + L) > 0. \quad \text{(A.4)}$$

To see this, let S be a real constant greater than all the magnitudes $A, B, \ldots, L$. Then take $R = 1 + \sqrt{2}S$. If $r > R$, then $r > 1 + \sqrt{2}S$, and so $1 - \frac{\sqrt{2}S}{r-1} > 0$. Since $r > R > 1$, we have that $\frac{1}{r} + \frac{1}{r^2} + \cdots + \frac{1}{r^n} < \frac{1}{r-1}$. It follows then that

$$r^n(1 - \sqrt{2}S(\frac{1}{r} + \frac{1}{r^2} + \cdots + \frac{1}{r^n})) > 0$$

hence

$$r^n - \sqrt{2}S(r^{n-1} + \cdots + 1) > 0$$

and so

$$r^n - \sqrt{2}(Ar^{n-1} + Br^{n-2} + \cdots + L) > 0.$$

Second, the circumference of a circle of radius $r > R$ will consist of $2n$ arcs, inside of which T takes alternately positive and negative values as in Figure A.1.

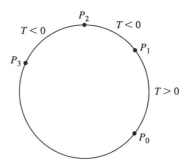

Figure A.1. T on a Circle of Radius r

To see this, consider $\theta = 4\pi/n$, and on a circle of radius $r > R$ consider $4n$ points $P_0, P_1, \ldots, P_{4n-1}$ with respective arguments $\theta, 3\theta, \ldots, (8n-3)\theta, (8n-1)\theta$.

The arguments corresponding to P_{2k} and P_{2k+1} are $\phi = (4k+1)\frac{\pi}{4n}$ and $\phi' = (4k+2)\frac{\pi}{4n}$. It follows that

$$\cos(n\phi) = (-1)^k \frac{1}{\sqrt{2}} \text{ and } \cos(n\phi') = (-1)^{k+1} \frac{1}{\sqrt{2}}.$$

Multiplying the corresponding values of T by $(-1)^k$ and $(-1)^{k+1}$ respectively we obtain

$$(-1)^k T = \frac{r^n}{\sqrt{2}} + (-1)^k A r^{n-1} \cos((n-1)\phi + \alpha) + \cdots + (-1)^k L \cos \lambda$$

$$(-1)^{k+1} T = \frac{r^n}{\sqrt{2}} + (-1)^{k+1} A r^{n-1} \cos((n-1)\phi' + \alpha)$$
$$+ \cdots + (-1)^{k+1} L \cos \lambda.$$

Replacing $(-1)^k \cos((n-1)\phi + \alpha), \ldots, (-1)^k \cos \lambda$, and $(-1)^{k+1} \cos((n-1)\phi' + \alpha), \ldots, (-1)^{k+1} \cos \lambda$ by -1 we get the following inequalities:

$$(-1)^k T \geq \frac{r^n}{\sqrt{2}} - A R^{n-1} - \cdots - L,$$

$$(-1)^{k+1} T \geq \frac{r^n}{\sqrt{2}} - A R^{n-1} - \cdots - L.$$

The right sides of both of these inequalities are positive by the choice of R, thus proving the original statement.

Now, since T varies continuously with ϕ, it follows that it will be zero at least $2n$ times on the circle. Suppose there are $2n$ zeros at the points $Q_0, Q_1, \ldots, Q_{2n-1}$, which have arguments between $\theta, 3\theta, 5\theta, 7\theta, \ldots, (8n-3)\theta, (8n-1)\theta$ respectively. Let $\zeta = \tan(\phi/2)$, and then

$$\cos \phi = \frac{1-\zeta^2}{1+\zeta^2}, \quad \sin \phi = \frac{2\zeta}{1+\zeta^2}.$$

Now let $z = r(\frac{1-\zeta^2}{1+\zeta^2} + i \frac{2\zeta}{1+\zeta^2})$. Then if we consider the original polynomial $f(z)$, its real part T becomes

$$T = \frac{p_{2n}(\zeta)}{(1+\zeta^2)^n}.$$

Here $p_{2n}(\zeta)$ is a real polynomial of degree no higher than $2n$. Since this polynomial has $2n$ zeros, it must be of degree exactly $2n$, and there can be no other roots. Therefore, $Q_0, Q_1, \ldots, Q_{2n-1}$ are the only points on this circle where T, vanishes and hence the positive and negative values of T alternate on this circle, as pictured in Figure A.1.

Thirdly, the values of the imaginary part U are positive at $Q_0, Q_2, \ldots, Q_{2n-2}$ and negative at $Q_1, Q_3, \ldots, Q_{2n-1}$. The argument ϕ of Q_k lies between $(4k+1)\pi/4n$ and $(4k+3)\pi/4n$. It follows that $(-1)^k \sin(n\phi) \geq 1/\sqrt{2}$. Multiplying U by $(-1)^k$ and replacing $(-1)^k \sin((n-1)\phi + \alpha), \ldots, (-1)^k \sin \lambda$ by -1 we have the inequality

$$(-1)^k U \geq r^n (-1)^k \sin(n\phi) - A R^{n-1} - \cdots - L$$

and hence also the inequality

$$(-1)^k U \geq \frac{r^n}{\sqrt{2}} - A R^{n-1} - \cdots - L.$$

App. A. A Version of Gauss's Original Proof

The right-hand side is positive by the choice of R and therefore the sign of U at Q_k is $(-1)^k$.

Now choose a circle Γ of radius $r > R$. Its circumference is divided into $2n$ arcs by points $Q_0, \ldots, Q_{2n-1}$ between which the sign of T alternates. As we expand the radius of the circle Γ the arcs $Q_0 Q_1$; $Q_1 Q_2$; etc. sweep out $2n$ regions extending to infinity within which T has alternately positive and negative values and these regions are separated by curves on which $T = 0$. Borrowing from Gauss and Uspensky's treatment, call the regions outside of Γ where $T > 0$ **seas** and the regions where $T < 0$ **lands**. The curves where $T = 0$ are then **seashores**. The n seas and n lands extend themselves into the interior of Γ across the various arcs. Starting from Q_1, move along the seashore so that the land is always on the right heading inward. We must eventually exit Γ, and when we cross it again the land must still be on the right. If the circumference is followed counterclockwise, lands and seas alternate. It follows then that we must exit Γ heading outward at an even point Q_{2k}, that is at Q_2, or Q_4, etc. Thus there is a continuous path γ leading from Q_1 to some point Q_{2k}. On path γ, $T = 0$. At Q_1, $U < 0$, whereas at point Q_{2k}, $U > 0$. Since U is a continuous function along the path γ, at some point on γ it takes the value 0. At this point T, U are both zero, which gives the existence of a root.

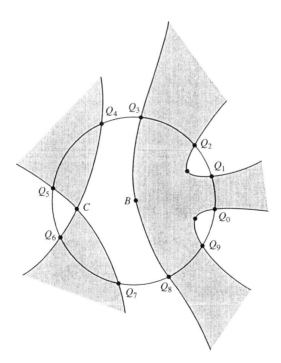

Figure A.2. Lands and Seas for a Certain Fifth-Degree Polynomial

The picture in Figure A.2 illustrates this for a special polynomial of degree 5 studied by Gauss. Starting from Q_1, we move along the shore to Q_2. At Q_1, $U < 0$, while at Q_2, $U > 0$. Along the shore $T = 0$, so there must be a point, marked A in the figure, where $U = 0$. This point represents a root. Similarly, the points B, C, D represent roots. The point C is a double root while the others are simple roots.

APPENDIX B Cauchy's Theorem Revisited

In Chapter 5 we gave our second proof of the Fundamental Theorem of Algebra based on Liouville's theorem in complex analysis. This in turn was based on Cauchy's theorem which says that if $f(z)$ is analytic in some domain U, then $\int_\gamma f(z)dz = 0$ for any closed continuously differentiable curve γ in U. At that time we gave a proof that depended on Green's theorem which in turn depended on the fact that $f'(z)$ is continuous. This was Cauchy's original proof. However, we mentioned that Goursat removed this restriction. In this section we return to this important theorem and give a proof of the Cauchy-Goursat theorem for star domains. In Appendix C, using some of the ideas we develop here, we will give three additional complex-analytic proofs of the Fundamental Theorem of Algebra.

A domain U in $\mathbb{C}$ is **path-connected** if any two points $z_1, z_2 \in U$ can be connected by a curve totally in U. Further, an open region $U \subset \mathbb{C}$ is path-connected if and only if for any two points $z_1, z_2 \in U$ there is **polygonal curve** between z_1, z_2. (By a polygonal curve we mean a curve made up of finitely many straight line segments. In the case of an open region this polygonal curve can be taken so that the linear pieces are parallel to the coordinate axes.)

We must extend somewhat the concept of line integrals. Let U be a region in $\mathbb{C}$ and let $\gamma_i : [a_i, b_i] \to U \subset \mathbb{C}, a_i \leq b_i, i = 1, 2$, be two curves in U such that $\gamma_1(b_1) = \gamma_2(a_2)$. Then the **sum** $\gamma = \gamma_1 + \gamma_2$ in U is defined as the curve

$$\gamma : [a_1, b_2 - a_2 + b_1] \to U$$

where

$$\gamma(t) = \gamma_1(t) \text{ if } t \in [a_1, b_1] \text{ and}$$
$$\gamma(t) = \gamma_2(t + a_2 - b_1) \text{ if } t \in [b_1, b_2 - a_2 + b_1].$$

Analogously, we can define the sum $\gamma_1 + \gamma_2 + \cdots + \gamma_m$ of finitely many curves in U. This addition of curves is associative.

A curve $\gamma : [a, b] \to \mathbb{C}, a \leq b$, is called **piecewise continuously differentiable** if there are finitely many continuously differentiable curves $\gamma_1, \ldots, \gamma_m$ such that γ is the sum $\gamma = \gamma_1 + \cdots + \gamma_m$, that is, there are points $a_1, a_2, \ldots, a_{m+1}$ with $a = a_1 < a_2 < \cdots < a_{m+1} = b$ such that γ_i is a curve with domain $|[a_i, a_{i+1}], 1 \leq i \leq m$, and is continuously differentiable. For example, each polygonal curve is piecewise continuously differentiable.

Now let U be an open region in $\mathbb{C}$ and $\gamma : [a, b] \to U$ a curve that is the sum $\gamma = \gamma_1 + \cdots + \gamma_m$ of finitely many continuously differentiable curves $\gamma_1, \ldots, \gamma_m$. Suppose $f : U \to \mathbb{C}$ is a continuous function. Then we define the line integral

$$\int_\gamma f(z)dz = \sum_{i=1}^m \int_{\gamma_i} f(z)dz.$$

The arc length of $\gamma = \gamma_1 + \cdots + \gamma_m$ as above is defined by $L(\gamma) = L(\gamma_1) + \cdots + L(\gamma_m)$, where as in Chapter 5 we have

$$L(\gamma_i) = \int_{\gamma_i} |dz| \text{ for } i = 1, \ldots, m.$$

If $|f(z)| \leq M$ on γ, then from Lemma 5.2.1 and the definition it follows that

$$\left| \int_\gamma f(z)dz \right| \leq ML(\gamma).$$

For the remainder of Appendices B and C we assume always that a curve is piecewise continuously differentiable as defined above.

Definition B.1
A subset $M \subset \mathbb{C}$ is **starlike** if there is a point $z_1 \in M$ such that for each $z \in M$, the line segment $\overline{z_1 z}$ is contained in M. z_1 is called a **center** of M. Clearly, starlike sets are path-connected.

$U \subset \mathbb{C}$ is a **star domain** if U is open and starlike.

EXAMPLE B.1
The following sets are all star domains.
  (a) $\mathbb{C}$ or $\mathbb{R}$ with center 0.
  (b) Any open circular region.

Appendix B: Cauchy's Theorem Revisited

(c) $H = \{z;\ \text{Im} z > 0\}$ with center i.
(d) $\mathbb{C}^- = \{z;\ z \notin \mathbb{R} \text{ or } z \in \mathbb{R},\ z > 0\}$. □

Definition B.2
Let $U \subset \mathbb{C}$ be open and $f : U \to \mathbb{C}$ be continuous. An analytic function $F : U \to \mathbb{C}$ is a **primitive function** for f in U if $F' = f$.

We obtain from Chapter 5 and the above definitions that if F is a primitve function for f in U, then

$$\int_\gamma f(z)dz = \int_\gamma F'(z)dz = F(\gamma(t_1)) - F(\gamma(t_0)),$$

where $\gamma(t_1), \gamma(t_0)$ are the endpoints of the curve γ. Hence, if a continuous function has a primitive function in U, it satisfies Cauchy's Theorem. Goursat's proof consists essentially in showing that any analytic function in a star domain has a primitive function. We first must relate integrability to the existence of primitive functions.

Definition B.3
Let $U \subset \mathbb{C}$ be open. Then $f : U \to \mathbb{C}$ is called **integrable** in U if f is continuous in U and if f has a primitive function in U.

Theorem B.1
(*Criteria for Integrability*) Let U be a domain and $f:U \to \mathbb{C}$ continuous in U. Then the following are equivalent:
   (1) f is integrable in U.
   (2) For each closed curve γ in U we have $\int_\gamma f(z)dz = 0$.
   If (2) is fulfilled, fix a point $z_1 \in U$ and for each $z \in U$ choose a curve γ from z_1 to z. Define

$$F(z) = \int_\gamma f(z)dz,\ z \in U.$$

Then $F(z)$ is a primitive function for $f(z)$ in U.

Proof
(1) clearly implies (2), so we must only show that (2) implies (1). Further, it is enough to show that $F(z)$ as defined in the statement is a primitive function of $f(z)$. Thus, we must show that $F'(z) = f(z)$ for all $z \in U$.
   Since $\int_\gamma f(z)dz = 0$ for each closed curve γ in U, the function $F(z)$ is well-defined (see Theorem 5.2.2). The proof is then the same as the proof of Theorem 5.2.2.

Appendix B: Cauchy's Theorem Revisited

$$F'(z) = \lim_{\Delta z \to 0} \frac{F(z + \Delta z) - F(z)}{\Delta z}$$

$$= \lim_{\Delta z \to 0} \frac{1}{\Delta z} \left(\int_{z_0}^{z+\Delta z} f(w)dw - \int_{z_0}^{z} f(w)dw \right)$$

$$= \lim_{\Delta z \to 0} \frac{1}{\Delta z} \int_{z}^{z+\Delta z} f(w)dw.$$

Since $f(z)$ is continuous, it can be shown (see the exercises in Chapter 5) that

$$\int_{z}^{z+\Delta z} f(w)dw = (f(z) + \epsilon)\Delta z,$$

where $\epsilon \to 0$ as $\Delta z \to 0$.

Therefore,

$$\lim_{\Delta z \to 0} \int_{z}^{z+\Delta z} f(w)dw = f(z),$$

and hence $F'(z) = f(z)$. ■

We now show that in a star domain U the integrability of $f(z)$ follows from the fact that $\int_{\partial \Delta} f(z)dz = 0$ for the boundary of any triangle Δ in U.

Theorem B.2
(*Triangle Criterion for Integrability*) Let U be a star domain with center z_0 and let $f : U \to \mathbb{C}$ be continuous. If

$$\int_{\partial \Delta} f(z)dz = 0$$

for the boundary $\partial \Delta$ of each triangle $\Delta \subset U$ with z_0 as a vertex, then $f(z)$ is integrable in U.

Further, the function

$$F(z) = \int_{\overline{z_0 z}} f(z)dz$$

is a primitive function for $f(z)$ in U, and in particular, $\int_{\gamma} f(z)dz = 0$ for any closed curve γ in U.

Proof
$F(z)$ is well-defined, since $\overline{z_1 z} \subset U$ for all $z \in U$. Let $c \in U$ be fixed and choose $z \in U$ such that the triangle Δ with vertices z_1, z, c is contained in U. We picture this in Figure B.1. Since $\int_{\partial \Delta} f(z)dz = 0$, we get that

$$F(z) = F(c) + \int_{\overline{cz}} f(w)dw.$$

Appendix B: Cauchy's Theorem Revisited

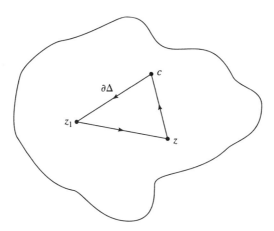

Figure B.1. Triangle Criterion

Then as in the proof of Theorem 5.2.2 and Theorem B.1 we get that F is a complex differentiable function with $F'(c) = f(c)$ for $c \in U$. ∎

We can now give the key criterion in Goursat's proof of the extended Cauchy theorem.

Theorem B.3
(*Integral Lemma of Goursat*) *Let $U \subset \mathbb{C}$ be open and $f: U \to \mathbb{C}$ be analytic. Let $\Delta \subset U$ be a triangle. Then*

$$\int_{\partial \Delta} f(z)\,dz = 0.$$

Proof
First we mention two elementary geometrical facts. Let Δ be any triangle and let L stand for length.
(1) $\text{Max}_{w,z \in \Delta}|w - z| \leq L(\partial \Delta)$.
(2) $L(\partial \Delta') = \frac{1}{2} L(\partial \Delta)$ for each of the four congruent subtriangles Δ' formed by drawing the line segments connecting the midpoints of the sides of Δ. These subtriangles are pictured in Figure B.2. ∎

Now define $v(\Delta) = \int_{\partial \Delta} f(z)\,dz$. We draw the line segments connecting the midpoints of the sides of Δ to get four congruent subtriangles Δ_v, $v = 1, 2, 3, 4$ as in Figure B.2. We then have

$$v(\Delta) = \sum_{v=1}^{4} \int_{\partial \Delta} f(z)\,dz = \sum_{v=1}^{4} v(\Delta_v),$$

since the integrals over the connecting lines occur twice with different signs and thus cancel. We picture the situation in Figure B.3.

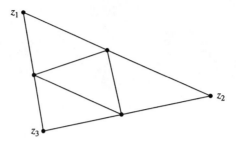

Figure B.2. Congruent Subtriangles

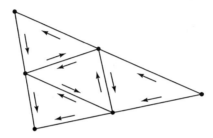

Figure B.3. Congruent Subtriangles

From the four integrals we choose one with maximal absolute value and let Δ^1 be the triangle for this chosen integral. Then

$$|v(\Delta)| \leq 4|v(\Delta^1)|.$$

Now we apply the same division and selection procedure to obtain a second subtriangle Δ^2 with

$$|v(\Delta)| \leq 4|v(\Delta^1)| \leq 4^2|v(\Delta^2)|.$$

If we continue in this manner we get a descending sequence

$$\Delta^1 \supset \Delta^2 \supset \cdots \supset \Delta^n \supset \cdots$$

of bounded closed triangles with

$$|v(\Delta)| \leq 4^n|v(\Delta^n)|, n = 1, 2, \ldots$$

From the second elementary geometrical remark made at the start of the proof we have that

$$L(\partial\Delta^n) = \frac{1}{2^n}L(\partial\Delta), n = 1, 2 \ldots$$

From the nested rectangles property in $\mathbb{R}^2$ we have that the intersection $\bigcap_{n=1}^{\infty}\Delta^n$ contains exactly one point $c \in \Delta$. Since $f(z)$ is analytic there is a continuous function $g : U \to \mathbb{C}$ with

$$f(z) = f(c) + f'(c)(z - c) + (z - c)g(z), z \in U \text{ with } g(c) = 0.$$

Appendix B: Cauchy's Theorem Revisited

From the existence of primitive functions we have that

$$\int_{\partial \Delta^n} f(c)dz = 0 \text{ and } \int_{\partial \Delta^n} f'(c)(z-c)dz = 0 \text{ for all } n \geq 1.$$

Hence we have that

$$v(\Delta^n) = \int_{\partial \Delta^n} (z-c)g(z)dz, \, n \geq 1.$$

This gives

$$v(\Delta^n) \leq \max_{z \in \partial \Delta^n}(|z-c||g(z)|)L(\partial \Delta^n).$$

From the first elementary geometrical fact mentioned, this inequality leads to

$$v(\Delta^n) \leq (L(\partial \Delta^n)\max_{z \in \partial \Delta^n}|g(z)|, \, n = 1, 2, \ldots$$

Altogether, we get that

$$|v(\Delta)| \leq 4^n |v(\Delta^n)| \leq L(\partial \Delta)^2 \max_{z \in \partial \Delta^n}|g(z)|, \, n = 1, 2, \ldots$$

Since $g(c) = 0$ and $g(z)$ is continuous at c, it follows that for each $\epsilon > 0$ there is a $\delta > 0$ with $|g(z)| \leq \epsilon$ on the ball $B_\delta(c)$ of radius δ centered on c. For this δ there is a positive integer n_0 such that $\Delta^n \subset B_\delta(c)$ for $n \geq n_0$. Hence $|g(z)| \leq \epsilon$ on the boundary of Δ^n for all $n \geq n_0$, and hence

$$|v(\Delta)| \leq (L(\partial \Delta))^2 \epsilon.$$

Since $L(\partial \Delta)$ is fixed and $\epsilon > 0$ is arbitrary, it follows that $v(\Delta) = 0$, proving the theorem.

We now have the general Cauchy's theorem for star domains. This is essentially the same theorem as was stated in Chapter Five. However in the proof we outlined there, we used Green's theorem, which assumed that the derivative $f'(z)$ was continuous. This turns out always to be the case. However, it is a consequence of the general Cauchy-Goursat theorem which just assumes that $f(z)$ is analytic.

Theorem B.4

(*Cauchy-Goursat Theorem for Star Domains*) *Let $U \subset \mathbb{C}$ be a star domain with center z_0 and let $f: U \to \mathbb{C}$ be analytic. Then $f(z)$ is integrable in U, and the function*

$$F(z) = \int_{\overline{z_0 z}} f(z) dz$$

is a primitive function for $f(z)$ in U. Further, it then follows that

$$\int_\gamma f(z)dz = 0$$

for any closed curve γ in U.

Proof

We have $\int_{\partial\Delta} f(z)dz = 0$ for the boundary $\partial\Delta$ of each triangle $\Delta \subset U$. The result then follows directly from the criterion for integrability. ∎

We give an example of Cauchy's theorem for star domains. Let $\mathbb{C}^- = \{z \in \mathbb{C}; z \notin \mathbb{R} \text{ or } z > 0\}$. $\mathbb{C}^-$ is a star domain with center 1. Let $f(z) = 1/z$. Then $f(z)$ is analytic in $\mathbb{C}^-$ and $\int_{\overline{1z}}(1/z)dz$ is a primitive function for $f(z)$.

Choose a curve from 1 to $z = re^{i\theta}$ as follows: take first the line segment on $\mathbb{R}$ from 1 to r and then the circular arc along $z = re^{i\phi}$ from r to z. Then

$$\int_{\overline{1,z}} \frac{dz}{z} = \int_1^r \frac{dt}{t} + \int_0^\theta \frac{ire^{it}}{re^{it}} dt = \log r + i\theta,$$

the main branch of the complex logarithm function in $\mathbb{C}^-$.

Using the Cauchy-Goursat theorem, Cauchy's integral formula and the Liouville theorem follow as in Chapter 5. From this we obtained our second proof of the Fundamental Theorem of Algebra. In Appendix C we use some other results in complex analysis to give three additional analytic proofs.

APPENDIX C

Three Additional Complex Analytic Proofs of the Fundamental Theorem of Algebra

The complex analysis proof for the Fundamental Theorem of Algebra that we gave in Chapter 5 was based on Liouville's theorem (Theorem 5.4.1). In this appendix we give three additional proofs not based on this result. We need two preliminary results. The first is a direct consequence of the Cauchy integral formula (Theorem 5.3.1).

Lemma C.1
(Mean Value Theorem and Inequality for Complex Integrals) Let $U \subset \mathbb{C}$ be open, $f : U \to \mathbb{C}$ be analytic, and let $B = B_r(z_0)$ be a circular disk of radius $r > 0$ centered on z_0 and contained in U. Then using the parametrization $\gamma(t) = z_0 + re^{it}, 0 \leq t \leq 2\pi$, for ∂B we get

$$f(z_0) = \frac{1}{2\pi} \int_0^{2\pi} f(z_0 + re^{it}) dt \qquad \text{(Mean Value Theorem)}$$

and

$$|f(z_0)| \leq max_{\partial B}|f| \qquad \text{(Mean Value Inequality)}.$$

The second preliminary result that we need is called the **growth lemma**.

Lemma C.2
(Growth Lemma) Let $p(z) = a_0 + a_1 z + \cdots + a_n z^n$ be a complex polynomial of degree $n \geq 1$ (so $a_n \neq 0$). Then there exists an $R \geq 1$ such that for all

195

$z \in \mathbb{C}$ with $|z| \geq R$ we have
$$\frac{1}{2}|a_n||z^n| \leq |p(z)| \leq 2|a_n||z|^n.$$
In particular, it follows that $|p(z)| \to \infty$ as $|z| \to \infty$.

Proof
Let $r(z) = |a_0| + |a_1||z| + \cdots + |a_{n-1}||z|^{n-1}$. Then
$$|a_n||z|^n - r(z) \leq |p(z)| \leq |a_n||z|^n + r(z)$$
by the triangle inequality. If $|z| \geq 1$, then we have $r(z) \leq k|z|^{n-1}$, where $k = \sum_{i=0}^{n-1} |a_i|$, because $|z|^i \leq |z|^{n-1}$ for $i < n$ and $|z| \geq 1$. Therefore we get the stated result with
$$R = \max\{1, 2k|a_n|^{-1}\}. \qquad \blacksquare$$

We can now give our three new proofs. Again, we want to prove that any nonconstant complex polynomial has a complex root.

Proof Seven
Let $p(z) = a_0 + a_1 z + \cdots + a_n z^n$ be a complex polynomial of degree $n \geq 1$ and assume that $p(z)$ has no zero in $\mathbb{C}$. As in Chapter 3 (see 3.4.1) let
$$\overline{p}(z) = \overline{a_0} + \overline{a_1} z + \cdots + \overline{a_n} z^n.$$
Then $\overline{p}(\overline{z}) = \overline{p(z)}$ for all $z \in \mathbb{C}$, and hence $g(z) = p(z)\overline{p}(z)$ has degree $2n$ and no zero in $\mathbb{C}$. Further, $g(x) = |p(x)|^2 > 0$ for $x \in \mathbb{R}$.

Since $\frac{1}{g(z)}$ is analytic in $\mathbb{C}$, we get from Cauchy's Theorem that for all $r > 0$
$$0 = \int_{-r}^{r} \frac{dx}{|p(x)|^2} + \int_{\gamma_r} \frac{dz}{g(z)}, \qquad (C.1)$$
where $\gamma_r(t) = re^{it}, 0 \leq t \leq \pi$, is the semicircle of radius r centered on 0. This is pictured in Figure C.1.

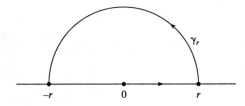

Figure C.1. Semicircle in Proof Seven

By the growth lemma we have
$$|g(z)|^{-1} \leq M|z|^{-2n}$$

with $M = 2|a_n|^{-2}$ for $|z| \geq r$, r sufficiently large. Then for such r we get that

$$\left|\int_{\gamma_r} \frac{dz}{g(z)}\right| \leq \max_{\gamma_r} \left|\frac{1}{g(z)}\right| |\pi r| \leq \pi M r^{-(2n-1)},$$

that is,

$$\lim_{r \to \infty} \int_{\gamma_r} \frac{dz}{g(z)} = 0,$$

since $n \geq 1$. But then from (C.1) we must have

$$\lim_{r \to \infty} \int_{-r}^{r} \frac{dx}{|p(x)|^2} = 0.$$

But this is impossible, since $|p(x)|^2 > 0$. ∎

Proof Eight

Let $p(z) = a_0 + a_1 z + \cdots + a_n z^n$ be a complex polynomial of degree $n \geq 1$ and assume that $p(z)$ has no zero in $\mathbb{C}$. Then $|f(z)| = \frac{1}{p(z)}$ is analytic in $\mathbb{C}$, and hence by the mean value inequality we have

$$f(0) \leq \max_{\partial B_r} |f(z)|$$

for each circular disk $B_r(0)$, $r > 0$, centered on 0. By the growth lemma we have $\lim_{z \to \infty} |f(z)| = 0$, and hence $f(0) = 0$ contradicting $f(0) = p(0)^{-1} \neq 0$. ∎

Our final proof depends on the **maximum minimum modulus principle**, which we mentioned in section 5.5. This is a consequence again of the Cauchy integral formula. Here we repeat it and two important corollaries and then give proof nine.

Theorem C.1
(Maximum Modulus Principle) Suppose $f(z)$ is nonconstant and analytic in a domain U. Then every neighborhood in U of $z_0 \in U$ contains points z with $|f(z)| > |f(z_0)|$.

Corollary C.1
(Maximum Principle) Suppose $f(z)$ is analytic in a domain U and assume that there is a point $z_0 \in U$ that is a local maximum for $f(z)$, that is, there is a circular disk $B_\epsilon(z_0) \subset U$ with $|f(z)| \leq |f(z_0)|$ for all $z \in B_\epsilon(z_0)$. Then $f(z)$ is constant in U.

Corollary C.2
(Minimum Principle) Suppose $f(z)$ is analytic in a domain U and assume that there is a point $z_0 \in U$ that is a local minimum for $f(z)$, that is, there is a

circular disk $B_\epsilon(z_0) \subset U$ with $|f(z_0)| \leq |f(z)|$ for all $z \in B_\epsilon(z_0)$. Then either $f(z) = 0$ or $f(z)$ is constant in U.

We need one other fact for the ninth proof. Since a polynomial over a field can have only finitely many zeros (see Chapter 3) it follows that a nonconstant complex polynomial $p(z)$ can be nowhere locally constant. That is there cannot be an open region $U \subset \mathbb{C}$ with $p(z)$ constant on U, for if $p(z) = c$ for all $z \in U$, then the polynomial $g(z) = p(z) - c$ has infinitely many zeros.

Proof Nine

Let $p(z) = a_0 + a_1 z + \cdots + a_n z^n$ be a complex polynomial of degree $n \geq 1$. For each $r > 0$ let $U_r = B_r(0)$ the closed ball of radius r centered on 0. Each U_r is a compact set and since $p(z)$ is continuous, $|p(z)|$ has a minimum at some point $z_r \in U_r$. Since $|p(z)| \to \infty$ as $|z| \to \infty$, we can choose an R sufficiently large such that this minimum point z_R is in the interior of U_R. To see this choose an $s > 0$ so that $|p(z)| > |a_0| = |p(0)|$ for $|z| > s$. Choose $R > s$. Then on the boundary of U_R we have $|p(z)| > |p(0)|$, and hence the minimum on the compact set U_R must be in the interior.

Therefore, we have a $z_R \in$ Interior U_r with $|p(z_R)| \leq |p(z)|$ for all $z \in U_R$. Since z_R is in the interior, there exists an $\epsilon > 0$ such that $U_1 = B_\epsilon(z_R)$, the closed ball of radius ϵ centered on z_R, is contained in U_R. However, $p(z)$, being everywhere analytic, is analytic on U_1, and further, $|p(z_r)| \leq |p(z)|$ for all $z \in U_1$. From the minimum principle either $p(z_R) = 0$ or $p(z)$ is locally constant on U_1. As mentioned above, a nonconstant polynomial can be nowhere locally constant, and therefore $p(z_R) = 0$. ∎

APPENDIX D

Two More Topological Proofs of the Fundamental Theorem of Algebra

In this final appendix we present two more topological proofs depending on winding number. The first is really a version of the Brouwer degree proof given in Chapter 9, while the second is a more formal presentation of the simple winding number argument given at the beginning of Chapter 8.

Recall that a **loop** in the unit circle S^1 is a function $f : [0, 1] \to S^1$ with $f(0) = f(1)$. As we saw in Chapter 9, the fundamental group $\Pi_1(S^1)$ is infinite cyclic, and hence each loop in S^1 has associated with it an integer m called its **degree**, or **winding number**. This can also be described as $1/2\pi$ times the change in the argument of z as z moves around the image of f in S^1, which if f is regular is equal to $1/2\pi \int_{image(f)} \frac{dz}{z}$. If g is a homotopic loop in S^1, then f and g have the same degree, and if $h = fg$ then degree(h) = degree(f) + degree(g). The map $S^1 \to S^1$ given by $z \to z^n$ clearly has degree n.

The set $\mathbb{C} - \{0\}$ is homotopic to an annular region, which in turn can be deformed into S^1. Thus, any loop in $\mathbb{C} - \{0\}$ is homotopic to a loop on S^1 and thus has a degree. We call this the degree, or winding number, of the loop around 0. In general, if $z_0 \in \mathbb{C} - \{0\}$, the degree, or winding number, of a loop $f(t)$ around z_0 is the degree of the loop $g(t) = f(t) - z_0$. The comments on homotopy and addition of degrees all carry over to general loops in $\mathbb{C} - \{0\}$. Our proof ten of the Fundamental Theorem of Algebra involves the following important result.

Theorem D.1
(Rouche's Theorem) Let $f, g; [0,1] \to \mathbb{C}$ be loops such that $|g(t)| < |f(t)|$ for all t. Then f and $f + g$ have the same degree.

Proof
The conditions insure that both f and $f + g$ map on $\mathbb{C} - [0]$. Consider $H(t, u) = f(t) + ug(t)$. Then $H(t, u)$ is a homotopy in $\mathbb{C}$ from $H(t, 0) = f(t)$ to $H(t, 1) = f(t) + g(t)$. The conditions on $f(t), g(t)$ further guarantee that there is no choice of (t, u) with $H(t, u) = 0$. Hence $H(t, u)$ is a homotopy in $\mathbb{C} - \{0\}$. Therefore, f and $f + g$ are homotopic and hence have the same degree. ∎

Proof Ten
Let $p(z) = a_0 + a_1 z + \cdots + a_n z^n$ be a complex polynomial with $a_n \neq 0$. For each $r \geq 0$ let $p_r : S^1 \to \mathbb{C}$ by $p_r(z) = p(rz)$. Then for any r the map $H(z, u) = p(zru)$ is a homotopy from p_r to p_0 in $\mathbb{C}$. If the polynomial $p(z)$ were never zero, this would be a homotopy in $\mathbb{C} - \{0\}$ and hence p_r and p_0 would have the same degree. Now, p_0 is a constant so it has degree 0. We will show that there is an R such that p_R has degree n. It will follow that $n = 0$, and therefore $p(z)$ is a constant.

Suppose $R > \max(1, \sum_{k=0}^{n-1} \frac{|a_k|}{|a_n|})$. Then for all z with $|z| = R$ we have

$$\left|\sum_{k=0}^{n-1} a_k z^k\right| \leq \sum_{k=0}^{n-1} |a_k||z|^k \leq \left(\sum_{k=0}^{n-1} |a_k|\right)|z|^{n-1} < |a_n||z|^n.$$

It follows then from Rouche's theorem that p_R has the same degree as the function on S^1 sending z to $a_n R^n z^n$, which has degree n. Since p_R and p_0 have the same degree, it follows that $n = 0$ and therefore, if $p(z)$ has no roots, $p(z)$ is a constant. ∎

The final proof is similar to the above but without recourse to Rouche's theorem.

Proof Eleven
As before, let $p(z) = a_0 + a_1 z + \cdots + a_n z^n$ be a complex polynomial with $a_n \neq 0$. Without loss of generality we can assume that $a_n = 1$. For each $r \geq 0$ let

$$p_r(t) = \frac{|p(r)|}{p(r)} \frac{p(re^{2\pi i t})}{|p(re^{2\pi i t})|}. \tag{D.1}$$

If $p(z)$ has no roots, then each of the two fractions in (D.1) is well-defined and has absolute value 1. Therefore, $p_r(t) \in S^1$ for each $r \geq 0$ and each $t \in [0, 1]$. Now, $p_0(t)$ is the constant loop in S^1. The first fraction gives an initial point to the loop $p_r(t)$, while the second fraction projects the loop $p(re^{2\pi i t})$ in $\mathbb{C} - \{0\}$ onto S^1.

Now, let $H(t, u) = p_{ru}(t)$. This gives a homotopy in S^1 from $p_r(t)$ to $p_0(t)$, and hence for each r, both p_r and p_0 have degree 0. We show that there is a value R where $p_R(t)$ has degree n.

Additional Topological Proofs

Let $S = \max(1, |a_1|, \ldots, |a_{n-1}|)$ and let $R = (n+1)S$. Let $f(t) = Re^{2\pi i t}$ be the circle of radius R centered on 0 and let $q(z) = z^n$. Then $q(f(t)) = R^n e^{2\pi n i t}$ is a loop with winding number n on a circle of radius R around the origin. Consider the distance from $q(f(t))$ to $p(f(t))$ for any t. We have

$$|q(f(t)) - p(f(t))| = |f(t)^n - (f(t)^n + a_{n-1}f(t)^{n-1} + \cdots + a_0)|$$

$$\leq |a_{n-1}f(t)^{n-1}| + \cdots + |a_0|$$

$$\leq |a_{n-1}||f(t)^{n-1}| + \cdots + |a_0|$$

$$\leq |a_{n-1}|R^{n-1} + \cdots + |a_0|$$

$$\leq \frac{R}{n+1}R^{n-1} + \cdots + \frac{R}{n+1} \leq \frac{n}{n+1}R^n < R^n.$$

This implies that for each t, $p(f(t))$ lies in a disk of radius R^n centered on $q(f(t))$. In particular, then, there is a line segment from $p(f(t))$ to $q(f(t))$ that does not go through the origin. This is in essence a version of the fellow-traveler property mentioned in Chapter 8. This says that the function $K(t, u) = uq(f(t)) + (1-u)p(f(t))$ is never zero for $u \in [0, 1], t \in [0, 1]$. Let

$$H(t, u) = \frac{|K(0, u)|}{K(0, u)} \frac{K(t, u)}{|K(t, u)|}.$$

By a straightforward computation we have that $H(t, 0) = p_R(t)$, $H(t, 1)$ is the loop $r(t) = e^{2\pi i n t}$, which has winding number n, and $H(0, u) = 1$ for all u. Therefore, $H(t, u)$ is a homotopy in S^1 from $H(t, 0) = p_R(t)$ to $H(t, 1) = r(t)$. It follows that $p_R(t)$ has degree n. Since it also has degree 0, we must have that $n = 0$ and $p(z)$ is constant. ∎

Bibliography and References

1 Quoted References

[A]   L. Ahlfors, *Complex Analysis*, McGraw-Hill, 1966.
[G]   N. Gupta, *The Burnside Problem*, The American Mathematical Monthly, 1991.
[H-Y] J.G. Hocking and G.S. Young, *Topology*, Addison-Wesley, 1961.
[R]   R. Remmert, *Fundamentalsatz der Algebra*, Zahlen, by H.D. Ebbinghaus, H. Hermes, F. Hirzebruch, M. Koecher, K. Mainzer, J. Neukirch, A. Prestel, R. Remmert, Springer-Verlag, 1983.
[Ro] J.Rotman, *Theory of Groups*, W.C. Brown - 3rd Edition, 1988.
[U]   J.V. Uspensky, *Theory of Equations*, McGraw-Hill ,1948.

2 Suggestions for Other Reading

Complex Analysis

L. Ahlfors, *Complex Analysis*, McGraw-Hill, 1966
E.A. Grove and G.Ladas, *Introduction to Complex Variables*, Houghton-Mifflin, 1974.
R.Remmert, *Funktionentheorie 1 and 2*, Springer-Verlag, (1984,1991)

General Algebra and Galois Theory

M. Artin, *Algebra*, Prentice-Hall, 1991.
H. Edwards, *Galois Theory*, Springer-Verlag, 1984
J.B. Fraleigh, *A First Course in Abstract Algebra*, Addison-Wesley - 5th Edition -, 1993.
S. Lang, *Algebra*, Addison-Wesley, 1984.
P.J. McCarthy, *Algebraic Extensions of Fields*, Dover, 1976.
B.L. Van Der Waerden, *Modern Algebra*, Ungar Publishing, New York - third printing, 1964.

Topology and Algebraic Topology

J.G. Hocking and G.S. Young, *Topology*, Addison-Wesley, 1961.
J.L. Kelley, *General Topology*, Van Norstrand, 1955.
K.H. Mayer, *Algebraische Topologie*, Birkhauser, 1989.
W.S. Massey, *Algebraic Topology: an Introduction*, Harcourt, Brace and World, 1967.
J.R. Munkres, *Topology: A First Course*, Prentice-Hall, 1975.
R. Stoecker and H. Zieschang, *Algebraische Topologie*, B.G. Teubner, 1988.

Gauss' Original Proofs

C.F. Gauss, *First Proof*, doctoral thesis Helmstadt - (in - Werke,3,1-30), 1799.
C.F. Gauss, *Second Proof*, Comm. Soc. Goett. **3**, , 1814/15, 107-142
C.F. Gauss, *Third Proof*, Comm. Soc. Goett. 3 - (in - Werke,3), 1816, 59-64
C.F. Gauss, *Fourth Proof*, Abhand. der Ges. der Wiss. zu Goett. **4**, 1848/50, 3-34

Index

Abel 104, 124
abelian group 5, 154–159
abelianization 158
absolute value 11
adjoining a root 81–84
algebraic closure 80, 85
algebraicially closed field 85
algebraic extension 75
algebraic integer 95
analytic function 40, 40–73
archimedean property 9
argument 16
automorphism group 107
Betti number 157
boundary 168
boundary group 172
Brouwer degree 177
Brouwer fixed point theorem 178
Burnside problem 156

Cardano 3–4, 124
Cauchy 42, 63, 185
Cauchy integral formula 66–69
Caucy sequence property 9
Cauchy's estimate 69
Cauchy's Theorem 64–69
Cauchy-Goursat Theorem 64–69
Caucy-Riemann equations 41, 41–46
Cauchy-Scwarz inequality 142
chain group 172
characteristic of a ring 113
classification problem 146
closed set 145
commutative ring 5
commutator subgroup 158
compact 37, 148–149
completely regular space 147
Complex Analysis 36, 36–73

205

complex conjugate 11
complex function 36
complex numbers 10, 10–19
complex plane 12
complex polynomial 1, 29–31
Complex Variables 36, 36–73
conformal mapping 46–49
conjugate of a subgroup 108
conjugate of a polynomial 30
connected 37, 149
constant polynomial 21
constructible number 126–130
constructions geometric 126–130
continuous 38
contour integral 61–63
contractible space 161
cover 148
covering space 166
curve 46
cycle group 172
cyclic group 18, 155
D'Alembert 3
degree of an extension field 8
degree of a polynomial 21
Demoivre's Theorem 17–19
differentiable 39
differential form 55
direct product 154
division ring 130
division algebra 130
division algorithm 24
domain 37
double integral 52
elementary symmetric
    polynomial 89
entire function 40
Euclidean algorithm 25

Euclidean space 140
Euler's Magic Formula 16
Euler's identity 16
exact 55
extension field 6
extreme values theorem 32
factor group 110–111
factor ring 83
fellow-traveler property 135
field 5
finite group 86
free abelian group 155
free group 165
fundamental group 163–167
Fundamental Theorem of Algebra
    1
    proofs of 32–33, 70–71, 91–93,
        123–124, 135–136, 178–180,
        182–186, 196–197, 197, 198,
        200–201
Fundamental Theorem of
    Arithmetic 24
Fundamental Theorem of Galois
    Theory 119
Galois 104
Galois extension 112–115
Galois group 115–119
Galois group of a polynomial 117
Galois Theory 104–133
Gauss 3
generators 88
Girard 3
Goursat 64
Green's Theorem 57–59
ground field 75
group 86
growth lemma 195

Index

harmonic function 45–46
Hausdorf space 147
holomorphic 40
homeomorphism 146
homologous 172
homology group 172–176
homology theory 153, 166–179
homomorphism of groups 106
homomorphism of rings 10
homotopic 153, 159–160
homotopically equivalent 153, 160
homotopy invariant 153, 161
homotopy theory 153
ideal 82–84
image 110
imaginary unit 10
inner product 140
inner product space 140–142
integral domain 22
intermediate field 75
intermediate value theorem 9
irreducible polynomial 24
isomorphism of fields 106
isomorphism of groups 106
Symmetric Polynomials 89, 99–102
isomorphism of rings 10
isogonal 47
Jordan Curve Theorem 57
kernel 110
Lagrange 108
Lagrange's Theorem 108
Laplace's equation 45
least upper bound 9
Lindemann 126
line integral 52–61
linear polynomial 21

Liouvilles' Theorem 70–71
loop 161
magnification 48
maximal ideal 83
maximum principle 72
metric space 139–140
metrizable space 147, 148
minimal polynomial 94
modulus 11
Morera's Theorem 71
neighborhood 37
netsed intervals property 9
normal extension 112
normal space 147
normal subgroup 108
normed linear space 141
open ball 143
open base 148
open set 143, 145
ordered field 8
path 161
path-connected 163
permutation 86–88
piecewise continuously differentiable 185
point-set topology 136, 136–151
polar form 14
polynomial 21
primitive factor 189
primitive polynomial 94
primitive roots of unity 18
principal ideal 82
quadratic polynomial 21
quaternion algebra 131
quaternions 131
quotient group 110–111
quotient ring 83

real numbers 8–9
real polynomials 29–31
regular curve 46
relations 88
Riemann 42
Riemann Sphere 179
ring 5
ring of polynomials 21–22
roots of polynomials 23, 27–28
roots of unity 18
Rouche's Theorem 197
Ruffini 4
second countable space 148
separable extension 112
separation axioms 147
simple closed curve 56
simple extension 77
simple group 109
simplex 167–168
simplicial complex 168–170
simplicial decomposition 170
simply connected 163
skew field 130
solvable by radicals 125
solvable group 125
splits 84
splitting field 84–86
star domain 188
stereographic projection 178

subcover 148
subfield 6
Sylow subgroup 111
symmetric group 86–88
symmetric polynomial 89–91, 99–102
T_1-space 147
topology 145
Topology 134–181
topological invariant 152
topological space 145
torsion group 156
totally real field 130
transcendence of e and π 94–99
transcendental extension 75
transcendental number 8
triangulation 170
trichotomy law 8
unique factorization domain 24
UFD 24
unit 22
universal covering space 166
vector space 7
Wantzel 126
winding number 134
zero divior 22
zero of a polynomial 23
zero ploynomial 21

Undergraduate Texts in Mathematics

(*continued*)

Kemeny/Snell: Finite Markov Chains.
Kinsey: Topology of Surfaces.
Klambauer: Aspects of Calculus.
Lang: A First Course in Calculus. Fifth edition.
Lang: Calculus of Several Variables. Third edition.
Lang: Introduction to Linear Algebra. Second edition.
Lang: Linear Algebra. Third edition.
Lang: Undergraduate Algebra. Second edition.
Lang: Undergraduate Analysis.
Lax/Burstein/Lax: Calculus with Applications and Computing. Volume 1.
LeCuyer: College Mathematics with APL.
Lidl/Pilz: Applied Abstract Algebra.
Macki-Strauss: Introduction to Optimal Control Theory.
Malitz: Introduction to Mathematical Logic.
Marsden/Weinstein: Calculus I, II, III. Second edition.
Martin: The Foundations of Geometry and the Non-Euclidean Plane.
Martin: Transformation Geometry: An Introduction to Symmetry.
Millman/Parker: Geometry: A Metric Approach with Models. Second edition.
Moschovakis: Notes on Set Theory.
Owen: A First Course in the Mathematical Foundations of Thermodynamics.
Palka: An Introduction to Complex Function Theory.
Pedrick: A First Course in Analysis.
Peressini/Sullivan/Uhl: The Mathematics of Nonlinear Programming.
Prenowitz/Jantosciak: Join Geometries.
Priestley: Calculus: An Historical Approach.
Protter/Morrey: A First Course in Real Analysis. Second edition.
Protter/Morrey: Intermediate Calculus. Second edition.
Roman: An Introduction to Coding and Information Theory.
Ross: Elementary Analysis: The Theory of Calculus.
Samuel: Projective Geometry.
Readings in Mathematics.
Scharlau/Opolka: From Fermat to Minkowski.
Sethuraman: Rings, Fields, and Vector Spaces: An Approach to Geometric Constructability.
Sigler: Algebra.
Silverman/Tate: Rational Points on Elliptic Curves.
Simmonds: A Brief on Tensor Analysis. Second edition.
Singer/Thorpe: Lecture Notes on Elementary Topology and Geometry.
Smith: Linear Algebra. Second edition.
Smith: Primer of Modern Analysis. Second edition.
Stanton/White: Constructive Combinatorics.
Stillwell: Elements of Algebra: Geometry, Numbers, Equations.
Stillwell: Mathematics and Its History.
Strayer: Linear Programming and Its Applications.
Thorpe: Elementary Topics in Differential Geometry.
Toth: Glimpses of Algebra and Geometry.
Troutman: Variational Calculus and Optimal Control. Second edition.
Valenza: Linear Algebra: An Introduction to Abstract Mathematics.
Whyburn/Duda: Dynamic Topology.
Wilson: Much Ado About Calculus.